AF454353

ISBN 978-90-831434-2-2

9 789083 143422 >

Published by R&A. ISBN 978-90-831434-2-2. Official price: 69.69 (in your favourite currency)

Cover image: (left) Plant leaves referenced from an AI-generated image and (right) chloroplasts referenced from electron microscope pictures, both inspired by the style of Ecuadorian painter Oswaldo Guayasamín. Illustration by Arac/Renske Wierda.

Abstract

Plants are sensitive to changing light conditions, and consequently, they have evolved a number of different mechanisms to adapt to such changes. One particular mechanism involves the motion of chloroplasts. Even though the motion of chloroplasts has been studied for over a century, the physics behind their light-dependent collective motion remains elusive. These organelles can move in different ways depending on the intensity of light they are exposed to. Chloroplasts move towards the sides of the plant cell under high-intensity light, while they accumulate at the top and bottom of the cell under low-intensity light. The myriad of different dynamic phase transitions seen in this system makes it fascinating to study, as it can give us new insights into the dynamics of dense biological active matter. Here, we study the active motion of chloroplast under dim-light conditions and during the transition from dim-to-bright light, using numerical simulations. For the dim-light adapted state we observe burst-like re-arrangements and show that these dynamics resemble those of colloidal systems close to the glass transition. We provide a model to investigate the dependence of the stability of this configuration on various system parameters. Our study of the dim-to-bright light transition shows that light gradients drive chloroplast motion and that an effective attraction between the particles due to the light leads to clustering. However, other physical effects are needed to reproduce the clustering dynamics seen in experiments. Nonetheless, the proposed agent-based model provides a solid basis to build a unified model for chloroplast motion that could be used to study the different dynamic phase transitions in this biological active matter system.

Contents

1 Introduction

Plants are some of the most important living organisms on the planet since they produce oxygen [1], an important metabolite for other life forms on earth, as a byproduct of photosynthesis. Plant cells are able to convert light into chemical energy via photosynthesis because they have organelles, sub-units inside their cells, called chloroplasts. Inside the chloroplasts are membrane-bound compartments called thylakoids that contain photosystems. Photosystems are light-capturing protein complexes that use light as fuel to generate the chemical energy needed to turn carbon dioxide into glucose while expelling oxygen. Plants are sensitive to different light intensities as light is their main energy source. On the one hand, exposure to high-intensity light can cause photodamage to the photosynthetic apparatus inside the plant cell. On the other hand, exposure to very low-intensity light diminishes the plants photosynthetic performance. In both cases the effect will be detrimental to the organism [2]. As plants cannot rapidly move when the light around them changes, they have developed an internal mechanism to control the amount of light the chloroplasts are exposed to: the chloroplasts themselves move in different ways depending on the intensity (and colour)[1] of the light [5–7].

The motion of chloroplasts has been studied since the second half of the 19th century [3, 9, 10], however a deeper understanding of the purpose of and mechanism behind this motion has only been developed recently [4–7, 11–14]. Fig. 1a shows a schematic drawing of the different types of motion that chloroplasts display. Chloroplasts accumulate at the top and bottom of the cell, the areas where the light intensity is highest, when they are exposed to low-intensity light in order to maximise light absorption and enhance photosynthesis [7]; this is called the **accumulation response**. After spreading out over the cell, the chloroplasts are in a **dim-light adapted state** [8], see Fig. 1b. When the plant cell is exposed to high-intensity light, chloroplasts move towards the the sides of the cell, away from the irradiated area, where they move around in concentric circles[2] (see Fig. 1a and c). This is called the **avoidance response** and it minimises light

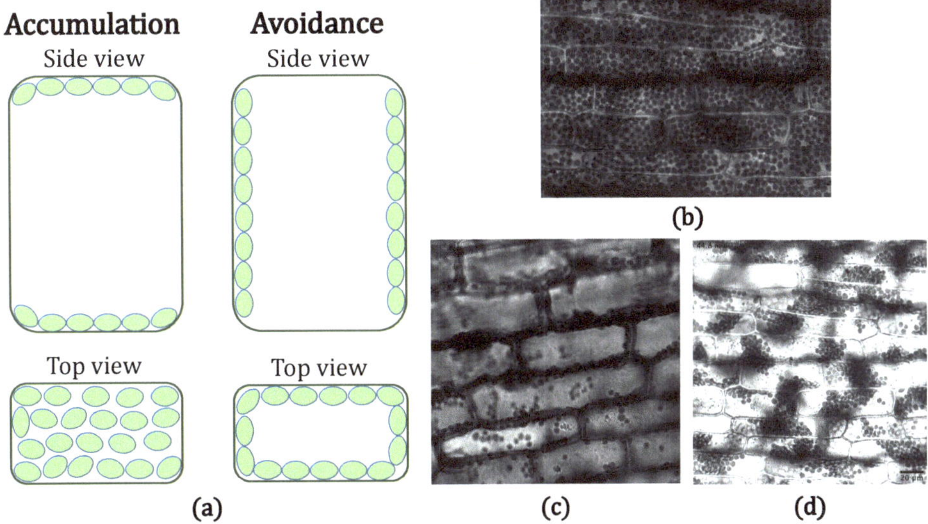

Fig. 1: **(a)** Schematic drawing of the motion of chloroplasts under different light intensities. Inspired by Fig. 1 from [8]. **(b)** Microscopic image of chloroplasts inside *Elodea densa* cells after prolonged exposure to weak intensity light. The chloroplasts are in the dim-light adapted state. **(c)** Microscopic image of chloroplasts inside *Elodea densa* cells during the avoidance response. The plant cell is exposed to high-intensity light for a prolonged period of time. **(d)** Microscopic image of chloroplasts inside *Elodea densa* cells during the dim-to-bright light transition. The plant cell is exposed to high-intensity light for a short period of time. The author thanks Nico Schramma for providing images (b-d).

---

[1]In this work we will not focus on the effect of different colours of light on chloroplasts motion. Different plants are sensitive to different colours of light. The effective wavelength for chloroplast motion is blue light, but red light can also be effective in some plant species [3, 4].

[2]Why the chloroplasts move around the edges of the cell in concentric circles is unknown.

absorption and reduces photodamage [5, 8]. Whenever there is a transition between low and high-intensity light, chloroplasts show yet another type of motion. In the waterplant *Elodea densa* we see that, as soon as high-intensity light is turned on, the chloroplasts start to cluster and these clusters move around in the cell[3], see Fig. 1d. We call this the **dim-to-bright light transition**.

The goal of this thesis is to investigate the dynamics of chloroplasts in the dim-light adapted state and during the dim-to-bright light transition. This is done through numerical simulations with two separate models for the different types of dynamics. This thesis is structured as follows: we start by disucssing the biochemical mechanism behind chloroplasts motion in Section 2. Since the mathematical models used in this work are based on experimental measurements we discuss these for both types of dynamics in Sections 3.1 and 4.1 respectively. We model the dynamics of the dim-light adapted chloroplasts using the overdamped Langevin equation for a Brownian particle in a harmonic potential that shifts when a critical displacement is exceeded. The details of the model are discussed in depth in Section 3.2 and the results can be found in Section 3.3. To model the dynamics of chloroplasts during the dim-to-bright light transition we use an agent-based Brownian dynamics model which we discuss in depth in Section 4.2. We use this model to test our hypotheses that light gradients drive chloroplast motion and that the clustering of chloroplasts is driven by an effective attraction between the particles due to the light. These results are discussed in Section 4.4.

Our research shines a light on the various dynamic phase transitions this system can undergo, which can also be found in active glasses and dense active matter [15, 16]. The agent-based Brownian dynamics model proposed in this work provides a basis to build a model for all chloroplasts motion. Consequently, this study provides a solid basis for future research into the different types of dynamics of chloroplasts. The transition from the accumulation to the avoidance response could be studied by combining such a model with experimental measurements. This will lead to a better understanding of the dynamics of dense biological active matter.

---

[3]This motion can be different in different plant species and cell geometries. The colour of the light also seems to have an effect on the chloroplast dynamics. During experiments we observed that the chloroplast clusters seem to move inside the cell in the late afternoon under yellow/red light. However, the exact effect of different colours of light is unknown.

2 Biochemical Mechanism behind Chloroplast Motion

Although the light-induced motion of chloroplast has been studied for a long time [3, 9, 10], the exact mechanism behind chloroplast motion is still being debated amongst biologists. However, a consensus has largely been reached that this mechanisms involves actin filaments [8, 17–21]. These are polymers consisting of sub-units of monomeric G-actin. Chloroplasts are attached to the plasma membrane, the membrane that lines the plant cell itself, through multiple proteins. Each chloroplast is surrounded by a specific type of actin filaments called **cp-actin** (**c**hloroplast actin) [17–19]. The amount of cp-actin that is bound to a chloroplast is controlled by a protein called CHUP1 (**Ch**loroplast **U**nusual **P**ositioning 1) [5, 22, 23] which is one of the membrane bound proteins on the chloroplasts outer membrane, the membrane that surrounds the chloroplast.

CHUP1 is responsible for polymerising cp-actin depending on signals received by two blue light photorecept-ors [4] from the phototropin family, phototropin 1 (phot1) and phototropin 2 (phot2) [20, 24–26]. These photoreceptors are located on the plasma membrane and phot2 specifically is also located on the chloroplasts outer membrane. Whenever a signal from one of the photoreceptors is received, CHUP1 is re-distributed on the chloroplast, leading to actin polymerising wherever the concentration of CHUP1 is highest [27]. This be-comes the leading side, i.e. front, of the chloroplast, which is pulled forward as the newly polymerising actin attaches to the chloroplast envelope and the plasma membrane through a molecule called THRUMIN1 [28], see Fig. 4 and 5. As such, the chloroplast slides forward by the newly polymerised actin filaments. If the position of the light source changes, CHUP1 is re-distributed on the chloroplast such that a new leading side is created [8, 25, 29]. This means that chloroplasts do not have a fixed front and back. Consequently, they do not need to turn in order move toward or away from light [30], see Fig. 2.

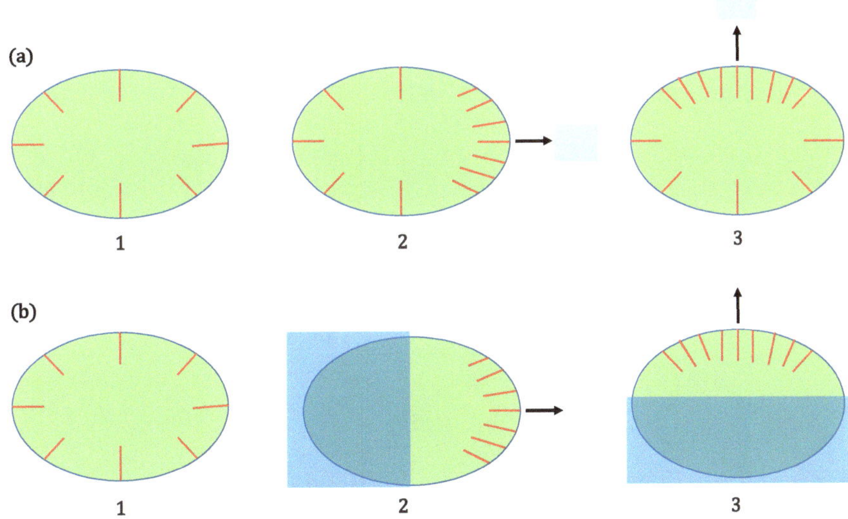

Fig. 2: Schematic drawing of actin-driven chloroplast motion. **(a)** Accumulation response: **1.** A chloroplast starts out with a homogenenous distribution of cp-actin (red bars) when there is no light shining on it. **2.** A beam of low-intensity light (light blue rectangle) illuminates an area to the side of the chloroplast, which leads to cp-actin polymerising on the side of the chloroplast closest to the light. Not all cp-actin on the trailing side of the chloroplast is depolymerised during this process. This leads to the chloroplast slowly moving towards the light. **3.** If the beam of light changes position a new leading side is created. **(b)** Avoidance response: **1.** A chloroplast starts out with a homogenenous distribution of cp-actin (red bars) when there is no light shining on it. **2.** Part of a chloroplast is illuminated with high-intensity light (blue rectangle), which leads to a polymerisation of cp-actin filaments on the side of the chloroplast away from the light and a depolimerisation on the side that is illuminated. This leads to the chloroplast rapidly moving away from the light. **3.** A different leading side is created if the light shines on a different side of the chloroplast. Figure inspired by Fig. 4 from [8].

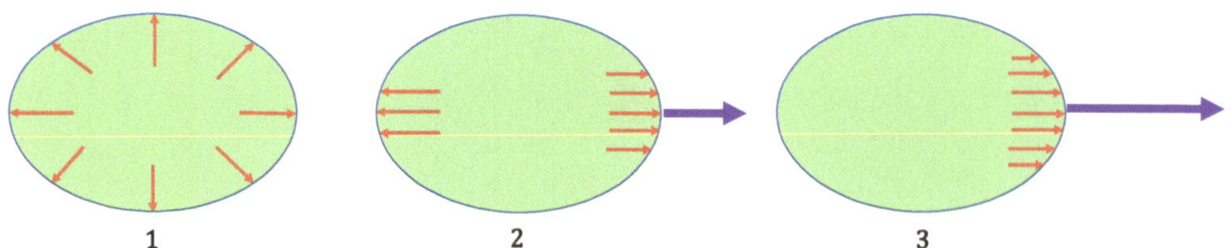

Fig. 3: Schematic drawing of actin-generated forces acting on a chloroplast. The red arrows depict the force generated by bound cp-actin filaments on the chloroplast. The purple arrows depict the net force acting on the chloroplast. **1.** If actin binding domains are evenly distributed the net-force equals zero and therefore there is no net movement. **2.** The cp-actin concentration on one side of the chloroplast is higher than on another. This leads to a net force pulling the chloroplast in the direction where the concentration is highest. This becomes the leading side of the chloroplast. **3.** The higher the concentration difference, the faster the chloroplast moves. Figure inspired by Fig. 4 from [8].

The actin concentration does not only determine the direction of motion of a chloroplast, but also its velocity. If the concentration difference on the front and back of the chloroplast is small, the chloroplast will move slowly. If the difference in concentration is large, the chloroplast will move rapidly [20, 21], see Fig. 3. Even though the force generation mechanism is the same for both the accumulation and avoidance response, they have different signal transduction pathways. In the following sections we will elaborate on these.

2.1   Accumulation response

The signal for the accumulation response is mediated by both phototropin photoreceptors, phot1 and phot2. The photoreceptors that control the accumulation response are located on the plasma membrane. Phot1 regulates accumulation over a broad range of blue-light intensities $(0.08 - 19.29\mathrm{Wm}^{-2})$ and phot2 over a smaller range of blue-light intensities $(0.38 - 3.85\mathrm{Wm}^{-2})$ [31]. Whenever phot1 and/or phot2 are activated by low-intensity light, a signal is sent that triggers the re-localisation of CHUP1 on the chloroplast envelope. This leads to actin polymerisation on the side of the chloroplast that is closest to the light, see Fig. 4. During accumulation, actin is not depolymerised in the back of the chloroplast, so the difference in actin concentration between the front and back of the chloroplast will be small [8, 25, 27, 29]. Therefore chloroplasts move slowly during the accumulation response.

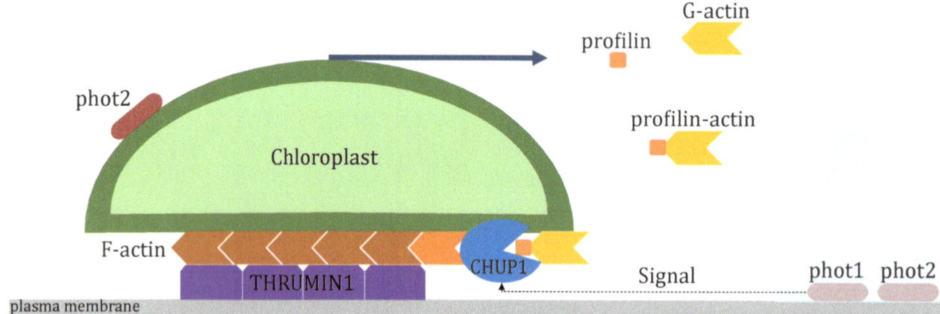

Fig. 4: Schematic representation of the molecular mechanism underlying chloroplast motion during the accumulation response. When low-intensity light shines on a plant cell (blue rectangle) a signal is released from photoreceptors phot1 and/or phot2 on the plasma membrane. This signal is received by a signal receptor associated with CHUP1. The activated CHUP1 starts polymerising actin monomers (profilin-actin) into cp-actin on the side of the chloroplast closest to the light. The cp-actin filaments are bundled by a protein complex called THRUMIN1 that is attached to both the plasma membrane and the chloroplast outer membrane [28]. This fixes the actin to the plasma membrane. Consequently, the chloroplast slides forward along the newly polymerised cp-actin filaments. For a more in depth description of the biochemical reactions involved in chloroplast motion see [8].

2.2 Avoidance response

The signal for the avoidance response is mediated only by phot2 located on the chloroplast outer envelope. Whenever this phot2 is activated, a signal is sent that triggers the re-distribution of CHUP1 to the side of the chloroplast facing away from the light. This leads to actin depolymerisation on the back of the chloroplast (close to the light) and polymerisation on the front (away from the light) [11, 27, 32], see Fig. 5. Consequently, there will be a very sharp actin gradient and the chloroplast will move rapidly away from the light.

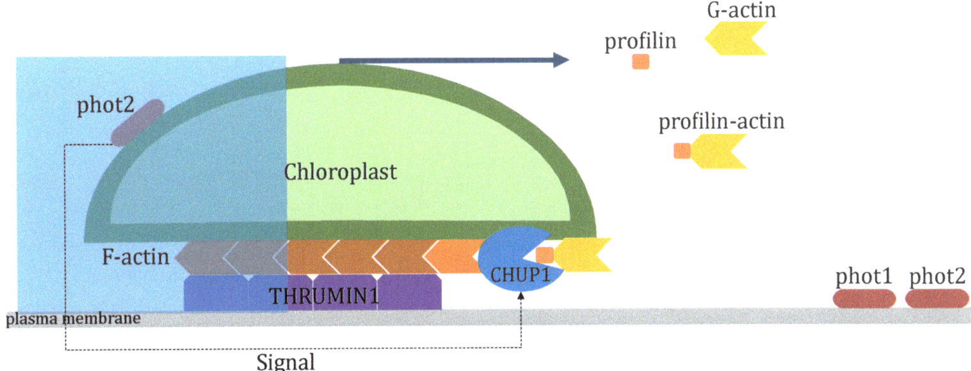

Fig. 5: Schematic representation of the molecular mechanism underlying chloroplast motion during the avoidance response. When high-intensity light shines on a plant cell (blue rectangle) a signal is released from photoreceptor phot2 on the chloroplast envelope. This signal is received by a signal receptor associated with CHUP1. The activated CHUP1 starts polymerising actin monomers (profilin-actin) into cp-actin on the side of the chloroplast away from the light. The cp-actin filaments are bundled by a protein complex called THRUMIN1 that is attached to both the plasma membrane and the chloroplast outer membrane [28]. This fixes the actin to the plasma membrane. Consequently, the chloroplast slides forward along the newly polymerised cp-actin filaments. For a more in depth description of the biochemical reactions involved in chloroplast motion see [8].

3   Dim-Light Environments

Having discussed the biochemical pathways that underlie chloroplast motion, we will now look into experimental data on the dynamics of these organelles in the dim-light adapted state. This means that the plant cell has already been exposed to low-intensity light for a prolonged period of time. These experimental observations will enable us to formulate a mathematical model of the dynamics. All experimental measurements where performed by Nico Schramma. A more in-depth description of the experiments can be found in [33]. The measurements were performed on the water plant *Elodea densa*. This plant was chosen because it has bilayered leaves, which makes imaging easier [34] and it is known to have strong and fast chloroplast motion [35].

3.1   Experimental measurements

The accumulation response is triggered when an *Elodea densa* cell is subjected to dim-light conditions (red-light of intensity $I = 7.2\,\mathrm{Wcm}^{-2}$). In these conditions the chloroplasts settle down on the periclinal walls, i.e. bottom of the cell, in a single layer. The 2D average packing fraction was measured to be approximately 71%, which suggests that individual chloroplasts are densely packed and their motion is highly dependent on the neighbouring chloroplasts. Analysis of individual chloroplast trajectories shows that, while many chloroplasts remain entrapped by their neighbours, a small fraction of them can escape their local confinement and travel along longer distances. Fig. 6a shows the displacement of these re-arranging chloroplasts together with their velocities. The movement of these particles shows a sudden displacement from the initial position which plateaus quickly at seemingly discrete distances. This shows that the slow chloroplasts transiently jump, but get caged rapidly again. Measurements confirm that re-arrangements of chloroplasts are connected to exceptionally high velocities compared to the confined chloroplasts. It is also apparent that the motion of multiple chloroplasts is temporarily correlated during these jump-like features in the displacement (Fig. 6a). This indicates that individual chloroplasts can break out of their confinement and drag their neighbours with them.

We also measure the mean square displacement (MSD) of the chloroplasts. The MSD is a measure of the deviation of the position of a particle with respect to a reference position over time. For a Markovian process, where the time-average and the ensemble-average are equal, the MSD is defined as:

$$MSD(\tau) = \overline{|\boldsymbol{r}(t+\tau) - \boldsymbol{r}(t)|^2} = \langle |\boldsymbol{r}(t) - \boldsymbol{r}_0|^2 \rangle. \tag{3.1}$$

The first term on the r.h.s is the time-average and the second term is the ensemble-average. For normal Brownian motion, during which a particle diffuses freely in space, the slope of the MSD scales linearly with Δt. Consequently, we have $\langle |\boldsymbol{r}(t)|^2 \rangle \propto D\Delta t$, with D the diffusion coefficient. The scaling of the MSD can also be super-diffusive, i.e. $\langle |\boldsymbol{r}(t)|^2 \rangle \propto D\Delta t^\alpha$ with $\alpha > 1$. If the MSD is sub-diffusive we have $\alpha < 1$. The MSD measured in the system of chloroplasts, see Fig. 6b, reveals a broad spectrum of power law exponents at small time scales ($\Delta t = 10 - 100\,\mathrm{s}$), ranging from sub-diffusive to super-diffusive trajectories. At long time scales of $\Delta t \gtrsim 5\,\mathrm{min}$, the MSD approaches an overall diffusive scaling. A few trajectories exhibit super-diffusive scaling, but most show sub-diffusive scaling laws at small times $\Delta t \lesssim 100\,\mathrm{s}$.

We also study the *self-part* of the Van Hove function, $\mathcal{G}_s(\Delta r, \Delta t)$. The Van Hove function, also referred to as the step-size distribution, is a dynamical correlation function that characterises the spatial and temporal distributions of pairs of particles. It is a probability distribution that gives us the probability of finding a particle at a certain position $\boldsymbol{r}$ and time t, given that one of the particles was at the origin at time $t = 0$. The self-part of the van Hove function describes the dynamics of a single particle. In Fig. 6c we see the results for our experiments. We note two different features of the Van Hove distribution function. Firstly, it is apparent that small displacements are governed by a transition of a non-Gaussian to a Gaussian step-size distribution for increasing time delays. This feature is commonly found in dynamics of (fractional) Brownian, non-Gaussian diffusive scalings and has been observed in organelles and RNA transport [36, 37]. Secondly, the step-size distribution has a heavy exponential tail for all lag times, which is an universal feature of particle motion close to the glass transition [38, 39]. We show that this is the case for our system by fitting the experimental results using a continuous-time random walk (CTRW) model (dashed line in

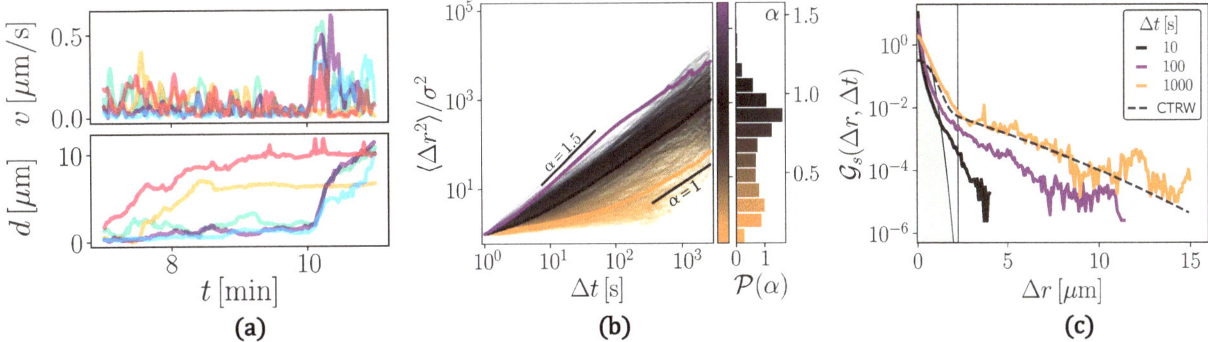

Fig. 6: **(a)** Velocity v and distance d from initial position for trajectories of chloroplasts which undergo strong re-arrangement. Different colours indicate different particles. Correlated motion of the sudden re-location is apparent from the trajectories. **(b)** Time-averaged MSD of all trajectories relative to the standard deviation. Colourmap represents the power-law scaling exponent α fitted on short time scales below $100\,$s. The exponent is widely distributed but peaks around a diffusive scaling regime ($\bar{\alpha} \approx 0.9$). While a few trajectories exhibit super-diffusive scaling, many trajectories show sub-diffusive scaling laws at small times $\Delta t \lesssim 100\,$s. Thick lines: ensemble averages over super-diffusive trajectories $\alpha \geq 1.1$ (purple), approximately diffusive trajectories $0.5 < \alpha < 1.1$ (black) and strongly sub-diffusive trajectories $\alpha \leq 0.5$ (orange). Notably on large time scales all regimes approach a diffusive scaling. **(c)** Self-part of the radial van Hove distribution function at different lag-times reveals non-Gaussian displacements on short length scales for short lag-times (compare with the gray area, representing a Gaussian function). For larger lag-times, the distributions approach a Gaussian, hence the comparison with a continuous time random walk (CTRW) model [38] (dashed line) improves. The author thanks Nico Schramma for providing all figures from [33].

Fig. 6c). This model was introduced by Montroll and Weiss [40] to effectively describe anomalous diffusion. It is a stochastic jump process with an arbitrary distribution of jump lengths and waiting times between jumps. The exponential decay of the step-size distribution (Fig. 6c) has been observed in other systems with various length scales. Some examples are: particle transport in human cells [41], motion of cells in confluent tissues [42], bacterial cytoplasm [43] and even systems like colloidal glasses [44] and super-cooled liquids [45].

With a high packing fraction ($\phi \approx 0.71$) the system of chloroplasts shows a remarkable similarity to (active) colloidal glasses. Studies of individual trajectories in colloidal glasses have led to the conclusion that the exponential decay of the step-size distribution at long time scales arises as a result of the coexistence of slow and fast particles [44]. This is called *dynamical heterogeneity*. This heterogeneity was also found in active systems and constitutes a hallmark for glasses [46]. Hence, our observation of the exponentially tailed step-size distribution indicates the existence of dynamical heterogeneity underlying chloroplast motion.

Considering all our previous arguments, we can say that here is a striking similarity between the dynamics of chloroplasts in the dim-light adapted state and many systems close to the glass transition. This implies that chloroplasts in this configuration obey similar physics to glassy systems. To study the effects of crowding and on chloroplast dynamics, we build a mathematical framework to model the co-existence of trapped and mobile particles.

3.2   Simulating dim-light chloroplast dynamics

The motion of chloroplasts is very complex, there are many effects that play a role: the activity of the particles, interactions between them, hydrodynamics, stochastic processes and many more. However, our aim is to make a simple model, taking only into account the stochastic nature of the movement and simplifying the interactions between the particles. We do not take into account hydrodynamic interactions or the activity of the particles. Since our goals is to describe a stochastic process, we use the Langevin equation. This equation describes the motion of a stochastic particle subject to random forces. Since the aforementioned experiments suggest that chloroplasts are at a state close to a glass-like transition, our goal is to infer their stability around this critical state.

3.2.1   The Langevin equation

The Langevin equation is a stochastic differential equation that describes the dynamics of a system of particles that is subjected to random and deterministic forces. The original equation put forward by Langevin [47] describes the motion of a Brownian particle. In one dimension we have

$$m\dot{v} = -\mu v + \xi(t) + f(t), \tag{3.2}$$

where m is the mass of the particle, v the velocity and μ is the drag coefficient from the hydrodynamic interactions between the particle and the fluid. $f(t)$ is an external force acting on the particle and $\xi(t)$ is the noise term that represents the effect of the collisions between the particle and the fluid. We assume we have Gaussian white noise with the following properties:

- $\langle \xi(t) \rangle = 0$,

- $\langle \xi(t)\xi(t') \rangle = 2D\delta_{ij}\delta(t - t')$,

where i and j are indices for the particles and D is the effective diffusion coefficient. This noise is in the limit of a-thermal noise, this means we ignore thermal fluctuations. How D is determined in our model will be explained in Section 3.3.

There are some simplification that can be made to this equation under certain circumstances. We estimate the mass of a chloroplasts to be approximately $m = 6.5 \cdot 10^{-14}$ kg, assuming the chloroplasts is a sphere with a radius $R = 2.5$ μm and it has the density of water. We use Stokes' law[4] [48] to estimate the drag coefficient a chloroplasts experiences. Using the dynamics viscosity of water at $20°C$, $\mu \approx 10^{-3}$ Pas, we find a drag coefficient $6\pi\mu R \approx 4.7 \cdot 10^{-8}$ kgs^{-1}. Taking the ratio of the mass and the drag coefficient gives us an inertial relaxation time scale τ_{inertial} for this system and we find $\tau_{\text{inertial}} \approx 10^{-6}$ s. Since this is very small, mass can be neglected and consequently the Langevin equation can be simplified. In such a case, where inertia can be neglected because of large damping or high viscosity, we have the Langevin equation for *overdamped dynamics*:

$$\dot{x} = f(x,t) + \xi(t). \tag{3.3}$$

Here we neglect all inertial forces and equate the velocity of the particle to all external forces $f(x,t)$ and the noise $\xi(t)$. Considering the value of τ_{inertial} we will assume we are always in the overdamped limit when modelling the dynamics of chloroplasts.

3.2.2   A single-particle toy model

We propose a one-dimensional Langevin-model of dim-light adapted chloroplasts in which the discrete jumps we see in experiments are determined by threshold phenomena. This model consists of a single bead attached to a wall with a spring with spring constant k, and a 'jump element' with critical value x_c, see Fig 7. In this system the particle will always feel a harmonic potential $V(x) = -kx^2$, but, if the particle moves past a critical position $\pm x_c$, the jump element will shift the potential and the position of the particle by a constant value $\pm\mathcal{B}$. The equation of motion for our system is

$$\dot{x} = \begin{cases} -\frac{k}{\gamma}(x - a(t)) + \sqrt{2D(t)}\xi(t) & \text{if} \quad |x - a(t)| < x_c, \\ -\frac{k}{\gamma}(x - a(t) \pm \mathcal{B}) + \sqrt{2D(t)}\xi(t) & \text{if} \quad |x - a(t)| \geq x_c. \end{cases} \tag{3.4}$$

Here $\xi(t)$ is a Gaussian white noise with $\langle \xi(t) \rangle = 0$, $\langle \xi(t)\xi(t') \rangle = \delta(t - t')$. $a(t) = (n_+(t) - n_-(t))\mathcal{B}$ is the baseline of the particle motion, which is proportional to the number of jumps up $n_+(t)$ minus the number of jumps down $n_-(t)$. This baseline is implemented to keep the particle inside the harmonic potential after every jump. Here the diffusion coefficient can be time-dependent. The effective spring constant k models the

---

[4]Gives the drag force exerted on spherical objects with small Reynolds number in a viscous fluid: $F_d = 6\pi\mu Rv$.

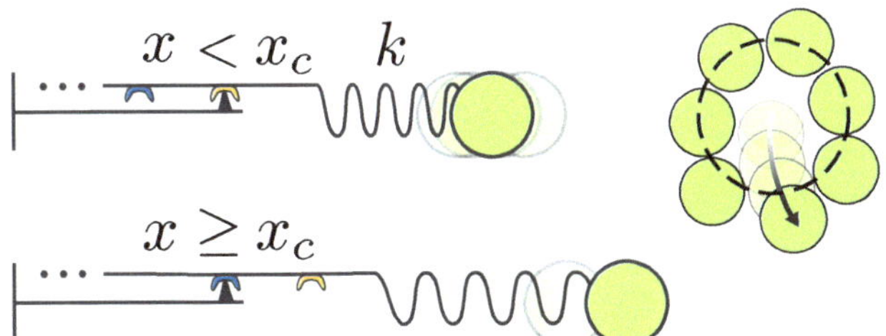

Fig. 7: Sketch of the model used for numerical simulations. A Brownian particle is attached to a wall with a spring with spring constant k. The spring is connected to a jump element with critical value x_c. If the displacement of the particle exceeds x_c the spring undergoes a sudden compression or extension and the potential is shifted by an overall constant $\mathcal{B}$. The author thanks Mazi Jalaal for providing this sketch of the model system.

confinement of a particle due to its neighbours [49, 50] and γ is a damping pre-factor. The jump element models the burst-like re-arrangements when a particle escapes from its confinement. The parameter x_c is a length scale whose value can be inferred from the waiting time between jumps. Short waiting times between jumps correspond to small values of x_c and vice versa.

The numerical simulations

Using the aforementioned model, we perform numerical simulations of the system. The numerical simulations were performed using the programming language Python. Since the simulations required a stochastic differential equation to be solved, we used the itoEuler algorithm from the package `sdeint`. The parameters of the model are shown in Table 1. All parameters in the simulations have arbitrary units.
We use the Euler-Maruyama algorithm for Ito equations, which requires our equations to be be written in Ito form:

$$dx = f(x,t)dt + g(x,t)dW. \tag{3.5}$$

Here $f(x,t)$ is a function for all the forces acting on the system, $g(x,t)$ is a function for the amplitude of the noise and dW are increments of Gaussian noise.

Parameter	Definition	Values used in simulations
x_0	Initial position of particle	0
T	Effective temperature[a]	1
$\frac{k}{\gamma}$	Ratio spring constant and damping pre-factor	$\frac{2}{18.849}$
D	Diffusion coefficient	$\sqrt{\frac{2k_B Tk}{\gamma}}$
dt	Integration time step	0.1
t_{tot}	Total integration time	10^5
x_c	Critical displacement	$[0, 4]$
$\mathcal{B}$	Shift of the harmonic potential	2

Table 1: Simulation parameters of the single-particle Langevin model for dim-light adapted chloroplasts.

[a]An active particle (like a chloroplast) seems to diffuse with an effective diffusion coefficient, i.e. effective temperature, on time scales longer than the persistence time of the particle.

For our model we have

$$f(x,t) = \begin{cases} -k(x - a(t)) & \text{if} \quad |x - a(t)| < x_c, \\ \frac{1}{dt}\frac{k}{\gamma}(-x + a(t) - x_c) - kx_c & \text{if} \quad x - a(t) \geq x_c, \\ \frac{1}{dt}\frac{k}{\gamma}(-x + a(t) + x_c) + kx_c & \text{if} \quad x - a(t) < x_c \end{cases} \tag{3.6}$$

and

$$g(x,t) = \sqrt{\frac{2k_B T k}{\gamma}} \tag{3.7}$$

In Eq. (3.6) the factor $\frac{1}{dt}$ is included to be able to integrate this equation properly with the chosen solver. The source code for this solver is available at [51]. In Eq. (3.7) we set $k_B = 1$ always. For each dataset we perform 100 runs of the simulation. We perform multiple runs in order to calculate the MSD and step-size distribution of the system. Parametric sweeps are done over x_c to elucidate how the values of this parameter affect the dynamics of the system. For all the results presented in the following section we assume that the diffusion coefficient is constant, unless explicitly stated otherwise.

3.3   Observation of glassy dynamics in chloroplasts

The simulation results provide us with the time series of the particle, which can be seen in Fig. 8 in purple. Using this data we calculate the velocity of the particle, shown in salmon pink in the same figure. The time series shows sudden jumps, which correspond to the particle exceeding the critical value x_c. These jumps are paired with a sharp increase in the velocity of the particle. We see a qualitative resemblance between the trajectories of the simulation and the experiments (see Fig. 6a): sudden jumps in the trajectory of the particle that are correlated with peaks in its velocity.

Furthermore, we investigate the MSD of the particle. For the simulations we calculate an ensemble-averaged MSD over all the runs, see Fig. 9a. The scaling of the MSD depends on the value of x_c. At long time scales of $\Delta t \gtrsim 10^1$ the MSD has a diffusive scaling for $x_c < 2$. As x_c increases the scaling changes from diffusive to sub-diffusive. These results are again qualitatively comparable to the the experimental MSD measurements (Fig. 6b). However, our model neglects the sub-diffusive dynamics at small time scales. From Fig. 9a we can conclude that our system shows similar caged dynamics to the chloroplasts. In our system, the caging is controlled by the threshold x_c. For small values of x_c the MSD has a diffusive scaling, which indicates that

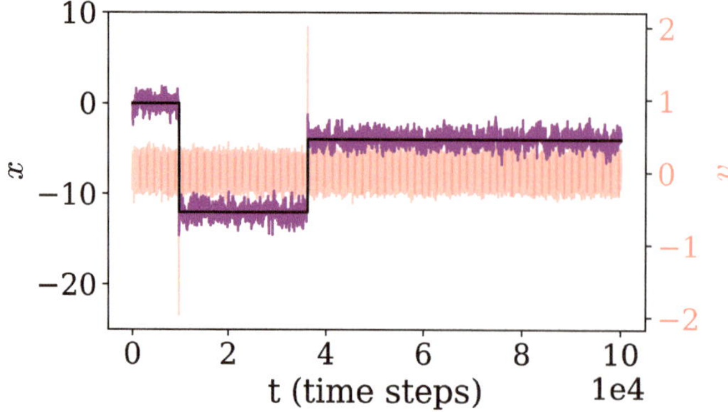

Fig. 8: Time series of a single particle attached to a wall by a spring and a jumping element with $x_c = 2.75$. The time series of the particle (purple) shows jumps whenever the particle exceeds x_c and these jumps are paired with a sudden peak in its velocity (salmon pink). Black line: indicates the base line of motion $a(t)$, see Eq. (3.4).

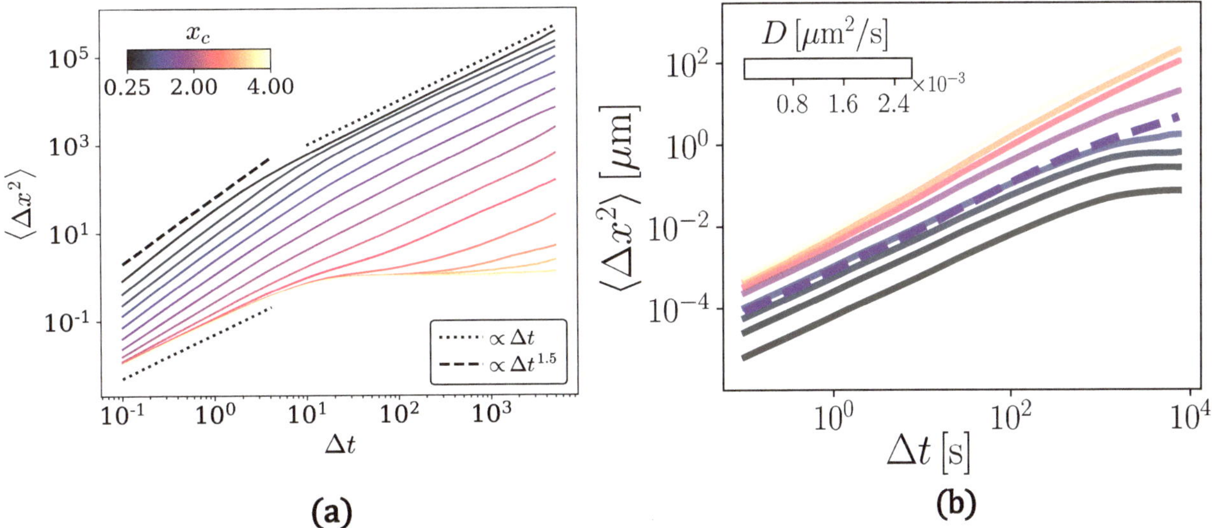

Fig. 9: **(a)** Ensemble-averaged MSD of a single particle attached to a wall by a spring and a jumping element with increasing x_c and constant diffusion coefficient. On large time scales the MSD is diffusive for small values of x_c, but as x_c increases the MSD plateaus and becomes sub-diffusive. This indicates a transition from free to caged motion. **(b)** Ensemble-averaged MSD of the same system but with diffusing diffusivity. As the diffusion coefficient increases, the MSD shows a transition from free to trapped motion. The author thanks Nico Schramma for providing this figure (b) from [33].

the particles are not strongly confined. For these values of x_c we see strongly pronounced jump dynamics in the time series. This is to be expected, since the Brownian particle is more likely to reach the critical displacement x_c whenever it is small. For larger values of x_c the system is strongly caged. The MSD has sub-diffusive scaling and the jump dynamics are far less pronounced since the system is less likely to exceed the critical value x_c[5]. This transition in the dynamics of the system suggests that chloroplasts are close to a critical point and can easily transition into a *liquid-like* state.

Lastly we look at the step-size distribution (van Hove function) of the system. Again we use the data from 100 runs, see Fig. 10. We see that the step-size distribution has a Gaussian center and broad exponential tails, which is qualitatively similar to the experimental results (Fig. 6c). However, our model in its current form does not account for the non-Gaussian to Gaussian transition at small displacements for increasing lag-times that we see in experiments. To account for this transitions we introduce **diffusing diffusivity** into our model. This means that we assume the diffusion coefficient is described by the square of an Ornstein-Uhlenbeck (OU) process [53, 54]. An OU process is a stochastic process that has a tendency to move back to the mean of the process. This attraction is stronger when the walker is farther away from the mean. It can be seen as a modified Wiener process, otherwise known as Brownian motion. An Ornstein-Uhlenbeck process is defined by Eq. (3.8).

$$\dot{Y}(t) = -\frac{Y}{\tau} + \sigma\eta(t), \tag{3.8}$$

where τ is the correlation time of the process and σ characterizes the amplitude of the fluctuations.

We assume that $D(t) = Y(t)^2$, with Y given by Eq. (3.8) We make this assumption based on the fluctuating and disordered nature of the chloroplast environment, and the fact that there is a wide range of diffusion coefficients which can change over time (see Section 3.1). This addition to our model describes the non-Gaussian to Gaussian transition of the van Hove function for small displacements with increasing time delays. The diffusing diffusivity model has been used previously to describe active processes in cytoskeletal dynamics [55] and the transport of RNA [37] or colloidal beads on membranes [36]. By adding diffusing

---

[5]The jump-time distribution in our model is, by definition, the mean first passage time of an Ornstein-Uhlenbeck process (following [52])

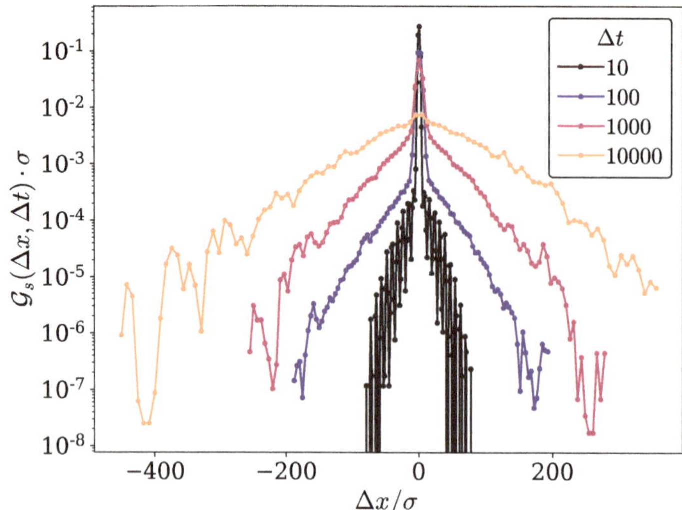

Fig. 10: Step-size distribution for different lag times Δt of a single particle attached to a wall by a spring and a jumping element with $x_c = 2$. The center of the distribution is Gaussian, while the tails are exponential. The distribution is normalised with respect to the standard deviation of the particle position σ.

diffusivity to our model, we find that the effective diffusivitiy $D = \sigma^2 \tau$ steers the caging dynamics in the same way that x_c does. This can be seen in the MSD of the trajectories (Fig. 9b). The system is strongly caged for small noise amplitudes σ, but as σ increases, the number of jumps becomes more pronounced. The implementation of this version of the model was done by Nico Schramma and the full set of results can be found in [33]. These results show that our model for the cage effects and dynamics of chloroplasts is enough to predict that either the critical size x_c or the activity σ can drive the system from a caged into a more liquid-like state.

In conclusion, our study shows that dim-light adapted chloroplasts resemble a system close to the glass transition. This means that the chloroplasts can go from a caged to a liquid-like state depending on their activity, which gives them the ability to respond to weak and bright light efficiently. We have seen that chloroplasts accumulate into a single layer at the bottom of the cell in response to weak light in order to increase their surface coverage inside the cell and therefore maximise photosynthetic efficiency. Moreover, chloroplasts must maintain the ability to avoid high-intensity light efficiently, without being hindered by other chloroplasts. Consequently, they must have a sufficiently loose packing and need to be able to increase their activity. In the following section we will take a closer look at the dynamics of chloroplasts during the dim-to-bright light transition.

4 Transition from Dim-to-Bright Light

Having studied the dynamics of dim-light adapted chloroplasts, we turn our attention to the dim-to-bright light transition. We will look into experimental measurements which, once again, will enable us to formulate a mathematical model of the dynamics. We use our model to test our hypotheses that light gradients drive chloroplast motion and that an effective attraction between the chloroplasts drives clustering under high-intensity light.

4.1 Experimental observations

As mentioned in Section 2, chloroplasts move when a force is applied through actin polymerisation [20]. The distribution of actin nucleation sites (CHUP1) on the chloroplast envelope was found to define the leading side of the chloroplast. By using a single optical fiber to locally impose a light stimulus on a chloroplast, Kong *et al.* found that the redistribution of CHUP1 on the chloroplasts envelope breaks the symmetry of the particle [27] (see Fig. 11a). From these experiments we learn two things: Firstly, the symmetry broken chloroplast shows a striking similarity to a Janus particle, since it has two sides that are chemically distinct and the biochemical reactions on one side of the particle drive its motion [56]. However, there is a key difference between Janus particles and chloroplasts. Janus particles are always symmetry broken, while chloroplasts are only symmetry broken when high-intensity light shines on them. Secondly, Kong *et al.* [27] also observe that a single chloroplast stays in the light beam when it is fully irradiated, while chloroplasts at the edges of the light source rapidly move away (see Fig. 11b). The difference between these chloroplasts is the fact that the fully irradiated chloroplast does not sense a light gradient, while the other chloroplasts do. This observation leads us to the hypothesis that a light gradient is key to chloroplast relocation and that this gradient drives the direction of motion.

We study the dynamics of a single chloroplasts in order to gain a better understanding of the response of

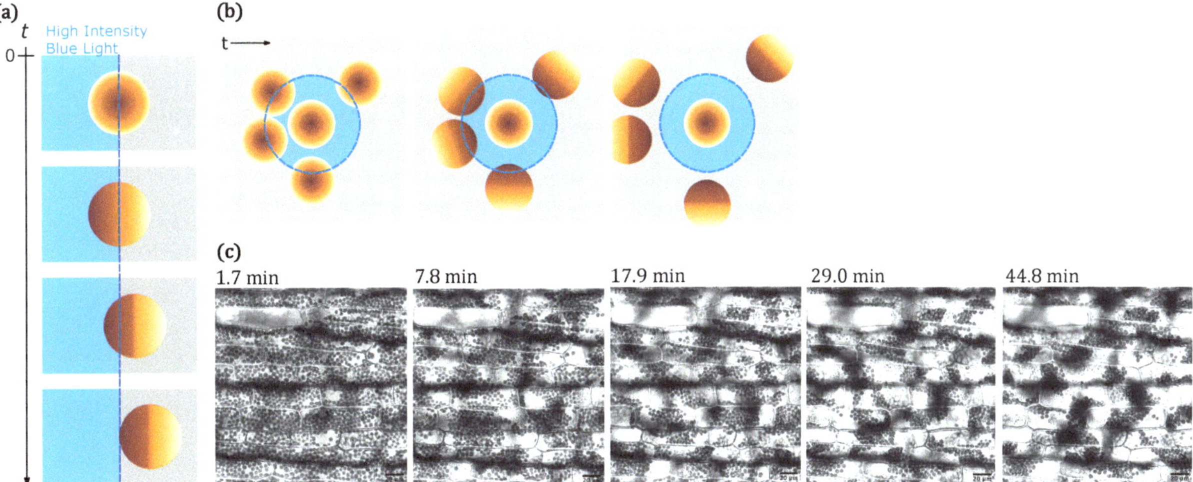

Fig. 11: **(a)** Schematic representation of the CHUP1 protein distribution (yellow) around a chloroplast (red) when it is irradiated locally by high-intensity blue light. **(b)** Schematic representation of the CHUP1 protein distribution (yellow) around a chloroplast (red) positioned inside a high-intensity light beam (blue circle). The chloroplasts at the edges of the beam move away from the light, but the chloroplast inside the beam stays in the same position and does not move. **(c)** Motion of chloroplasts in *Elodea densa* cells during the dim-to-bright light transition. The chloroplasts are in a dim-light adapted at $t = 1.7$ min. When the light is turned on the particles start to cluster. There is not one single cluster, but multiple clusters can form in the same cell and they can merge at later times. Occasionally, the clusters move inside the cell. Scale bar = 20 μm. Figures (a) and (b) are schematic representations of Fig. 1G and Fig.2B from Kong *et al.* [27]. The author thanks Nico Schramma for providing the experimental data for (c) and Renske Wierda for providing figures (a) and (b).

these particles to high-intensity light. Moreover, we are also interested in studying the collective aggregation of chloroplasts during the dim-to-bright light transition, see Fig. 11c. Active particle phase separation and aggregation is ubiquitous in many different system, from biological matter [57–59] to synthetic colloids [60–62]. These types of systems show Motility-Induced Phase Separation (MIPS), the phenomenon where a system of active particle phase separates due to the activity of the individual particles [63, 64]. The motion of chloroplasts during the dim-to-bright light transition could be due to MIPS. Studying the dynamics of this system could lead to a better understanding of phase separation and aggregation in dense biological active matter. Under dim-light conditions the chloroplasts have a high packing fraction. When the particles are irradiated with high-intensity light they quickly start to cluster, drastically decreasing the packing fraction (see Fig. 11c). The clusters are not always static, but they can move inside the plant cell in the case of *Elodea densa*. Small clusters can also detach from the main cluster and move freely. We hypothesize that the clustering of the chloroplasts is triggered by the shade neighbouring chloroplasts provide to each other when light shines on them. Since chloroplasts cast a shadow, this decreases the light intensity close to the particles and might create an effective attraction between them.

The distribution of CHUP1 during the avoidance response inspired us to model the dynamics of the system using the same type of model used for numerical simulations of chemotactic[6] Janus particles [56, 65]. Since we hypothesise that light gradients drive the motion of chloroplasts, we use a Langevin-model that couples the motion of an overdamped Brownian particle to the gradient of an external light field instead of a chemical gradient. We use our model to test if an effective attraction between the particles drives clustering dynamics by introducing a shadow effect between the model particles.

4.2 A multi-particle Langevin model

The basis for our mathematical model is a Langevin equation for an isotropic overdamped Brownian particle with center of mass coordinate r that couples to the gradient of a light field $I(r, t)$ [56]:

$$\dot{r} = \beta \nabla I(r(t), t) + \sqrt{2D}\xi(t). \tag{4.1}$$

Here β is the tactic coupling coefficient. If $\beta > 0$, the particle moves up the gradient, i.e. it is attracted by the light. If $\beta < 0$, the particle moves down the light gradient, i.e. it is repelled by the light. D is the diffusion coefficient of the particle, which we assume to be constant and ξ is a Gaussian white noise $\langle \xi(t) \rangle = 0$, $\langle \xi(t)\xi(t') \rangle = \delta(t - t')$.

This model captures the basic dynamics of a particles that responds to an external light gradient. However, the system we want to model has other interactions that need to be taken into account. Firstly, we need to add a repulsive interaction potential between particles, since chloroplasts have a physical size and can never fully overlap. Secondly, chloroplasts cast a shadow whenever bright light shines on them, which will lower the intensity of the light in the vicinity of each chloroplast. Consequently, this shadow effect must be implemented into the equation of the light field. Lastly, the particles need to be confined, since chloroplasts cannot move around freely in an infinite space. We will use a 2D model for chloroplast motion since the chloroplast are anchored to the plasma membrane [66] which allows for mostly two-dimensional motion. Taking all of this into account the final equation of motion for our system becomes:

$$\dot{r} = \beta \nabla I_{\text{total}}\left(r(t), \{r'_i(t)\}, t\right) - \nabla U_{cs}\left(r(t), \{r'_i(t)\}, t\right) + B(r(t)) + \sqrt{2D}\xi(t). \tag{4.2}$$

Here $r(t)$ is the position of the particle and $\{r'_i(t)\}$ are the positions of all other particles in the system. $I_{\text{total}}\left(r(t), \{r'_i(t)\}, t\right)$ is the external potential due to the light, $U_{cs}\left(r(t), \{r'_i(t)\}, t\right)$ is the interaction potential between particles and $B(r(t))$ is the boundary. In the following we will go through all the equations for the terms on the right hand side, starting with $I_{\text{total}}\left(r(t), \{r'_i(t)\}, t\right)$.

The term $I_{\text{total}}\left(r(t), \{r'_i(t)\}, t\right)$ consists of multiple functions that, together, generate the potential created by an external light source and the particles giving shade to each other. We implement this shading by subtracting a Gaussian that centered at the position of a second particle from the light field created by the external light source:

---

[6]Responding to an external chemical gradient.

$$I(\boldsymbol{r}(t), \{\boldsymbol{r}'_i(t)\}, t) = I_{\text{lamp}}(\boldsymbol{r}(t)) - \underbrace{C \exp\left(-\frac{|\boldsymbol{r}(t) - \sum_i \boldsymbol{r}'_i(t)|}{\lambda a^2}\right)}_{I_{\text{shadow}}}. \tag{4.3}$$

$I_{\text{lamp}}(\boldsymbol{r}(t))$ is the field created by an external source and I_{shadow} is the shadow created by the other particles in the system, with C the amplitude of the shadow and a the radius of the particle. The factor λ determines the width of the shadow. We set it to $\lambda = 0.32$ such that the shadow is just as wide as the particle. Increasing or decreasing λ will make the shadow wider or narrower respectively. In our simulations we set

$$I_{\text{lamp}}(\boldsymbol{r}(t)) = A, \tag{4.4}$$

with A the amplitude of the light, which is constant. This gives us a homogeneous light field. The shape of the potential of Eq. (4.3) is plotted in Fig. 12a.

We can also create an overall light gradient on the cell by introducing a function $S(\boldsymbol{r}, t)$ which we multiply with Eq. (4.3). This function can be changed as necessary to create any type of shadow over the cell. For example, to create a cell that is illuminated only when $x > 0$ we use the function:

$$S(\boldsymbol{r}, t) = \frac{1}{2}(1 + \tanh(\alpha x)). \tag{4.5}$$

We can increase or decrease the steepness of the light gradient at $x = 0$ by increasing or decreasing α respectively.

Lastly, we can create a light switch that changes the light field over time by introducing another function $L(t)$ that is also multiplied with Eq. (4.3). The function $L(t)$ can also be changed as necessary to create any type of changing light field[7]. We use Eq. (4.6) to turn the light on at time t_{on}. If we need a constant light field we set $t_{\text{on}} = 0$.

$$L(t) = H(t - t_{\text{on}}), \tag{4.6}$$

with $H(t)$ the Heaviside function.

Combining all of the equations mentioned above, we get the equation for the total light field:

$$I_{\text{total}}(\boldsymbol{r}(t), \{\boldsymbol{r}'_i(t)\}, t) = I(\boldsymbol{r}(t), \{\boldsymbol{r}'_i(t)\}, t)S(\boldsymbol{r}, t)L(t). \tag{4.7}$$

Next, we move on the the interaction potential $U_{cs}(\boldsymbol{r}(t), \{\boldsymbol{r}'_i(t)\}, t)$. We know that chloroplasts shade each other by partially overlapping in the z direction, i.e. one chloroplast can be on top of another. To model this behaviour in 2D we need an interaction potential that allows for strong repulsion on small length scales relative to the radius of the particle, but also enables partial overlap between the particles. This overlap does not model elastic compression of the particles, but the energetic cost of moving on top of another particle. We choose a core-softened potential [67] which consists of a short-range truncated Lennard-Jones term and a long-range Gaussian term centered around r_0 with depth u_0 and width c_0 [67], see Eq. (4.8).

$$U_{cs}(\boldsymbol{r}(t), \{\boldsymbol{r}'_i(t)\}, t) = 4\epsilon\left[\left(\frac{\sigma}{r - \{r'_i(t)\}}\right)^{12} - \left(\frac{\sigma}{r - \{r'_i(t)\}}\right)^{6}\right] + u_0 \exp\left(-\left(\frac{r - \{r'_i(t)\} - r_0}{c_0\sigma}\right)^2\right) \tag{4.8}$$

The short-range truncated Lennard-Jones term should give a strong repulsion and the long-range Gaussian term should provide an unstable repulsive region in the potential. This is necessary to be able generate an effective potential with overall attraction while keeping the potential repulsive on short length scales. The desired shape of the potential can be achieved with the correct values of the parameters ϵ, σ, u_0, c_0 and r_0, see Fig. 12b.

Whenever we have an external light field, the effective potential the particles are exposed to becomes a superposition of both the core-softened potential and the light potential. If the intensity of the light is strong

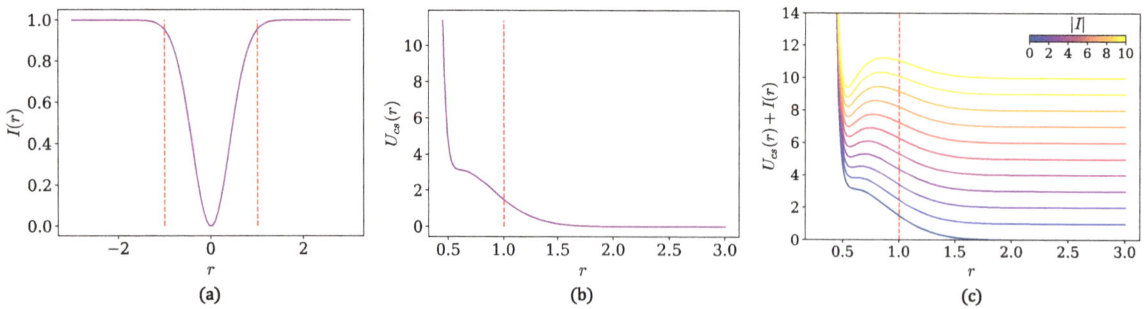

Fig. 12: **(a)** Light potential (Eq. (4.3)) felt by a particle with a neighbour at $x = 0$ with $A = 1$ and $C = A$.
(b) Core softened potential (Eq. (4.8)) with $\epsilon = 1$, $u_0 = 5\epsilon$, $c_0 = 1.3$ and $r_0 = 0.3$. **(c)** Effective potential.
For low light intensities the potential is fully repulsive, but as the intensity increases a global minimum starts
to appear. This indicates that there is an effective attraction between the particles. The strength of this
attraction depends on the light intensity. In all plots the red dashed line indicates the radius of the particle.

enough it can overcome the repulsion from the long-range Gaussian term of the core softened-potential. This
leads to an attraction between the particles that increases with the intensity of the light, see Fig. 12c.

Lastly, we have the boundary term $B(\boldsymbol{r})$ that keeps the particles confined to a cell. We implement a boundary
term that depends on the position $\boldsymbol{r}$ of a particle:

$$B(\boldsymbol{r}(t)) = 100 \left[H(-\boldsymbol{r}(t) - \boldsymbol{r}_{\max} + 1.1a)(-\boldsymbol{r}(t) - \boldsymbol{r}_{\max} + 1.1a)^2 - H(\boldsymbol{r}(t) - \boldsymbol{r}_{\max} + 1.1a)(\boldsymbol{r}_{\max} - \boldsymbol{r}(t) - 1.1a)^2 \right],$$
$$(4.9)$$

with $H(\boldsymbol{r})$ the Heaviside function and $\boldsymbol{r}_{\max}$ the boundaries of the cell. The factor $1.1a$ is introduced to keep
the particle inside the boundaries despite the noise in the system.

4.3   Details of the computational model

Having set up the mathematical structure of our model, we shift our focus to the numerical simulations of
the system.

In order to solve Eq. (4.2) numerically, we use the Rößler2010 order 1.0 strong Stochastic Runge-Kutta
algorithm SRI2 for Ito equations [68]. As for the jump-diffusion model, we use the solver from the `sdeint`
package which means our equations must be written in Ito form. The algorithm only provides the position of
the center of mass of each particle, but chloroplasts have a physical size and they measure the light around
them through photoreceptors on their surface (see Section 2). We know that the motion of chloroplasts is
generated by CHUP1 proteins binding actin at the boundary of the chloroplasts, see Fig. 11a. Consequently,
we only consider the light intensity at the boundary of the particle. Furthermore, we know that chloroplasts
move out of the light, even if only a small part of them is in the light. Consequently, our model needs to
give the same importance to large gradients. In order to give the model particles the correct direction of
motion away from the light, we need to calculate the force acting on the particle due to the light potential,
$\nabla I_{\text{total}}(\boldsymbol{r}(t), \{\boldsymbol{r}'_i(t)\}, t)$. The simplest way to do this is to take the average of the force over the boundary, but
this approach does not lead to the strong avoidance of light we see in experiments. Instead, as the particle
moves away from the light, the average force decreases significantly until it is no longer strong enough to
counteract the noise in the system. As a result the particle does not completely move out of the light.
To remedy this, we take the weighted average of the force, with as the weights the values of the intensity
squared, $|I_{\text{total}}(\boldsymbol{r}(t), \{\boldsymbol{r}'_i(t)\}, t)|^2$, at the boundary. The use of the weighted average makes sure that high
intensities have a higher impact compared to low intensities, which is what we require for fast motion away
from the light. We chose $|I_{\text{total}}(\boldsymbol{r}(t), \{\boldsymbol{r}'_i(t)\}, t)|^2$ as the weights because linear weights would not show the
same divergence at high intensities.

The Ito equations for this system thus are:

---

[7]For example, if we want the light to be turned on and off with a frequency ν we use $L(t) = \frac{1}{2}(\text{sign}(\sin(2\pi\nu t)) + 1)$.

$$f(\boldsymbol{r},t) = \beta \langle \nabla I_{\text{total}}(\boldsymbol{r}(t), \{\boldsymbol{r}'_i(t)\}, t) \rangle_\circ - \nabla U_{cs}(\boldsymbol{r}(t), \{\boldsymbol{r}'_i(t)\}, t) + B(\boldsymbol{r}(t)), \tag{4.10}$$

$$g(\boldsymbol{r},t) = \sqrt{2D}, \tag{4.11}$$

with

$$\langle \nabla I \rangle_\circ = \frac{\int_\mathcal{B} \nabla I(\boldsymbol{r}(\theta,t), \{\boldsymbol{r}'_i(t)\}, t) |I(\boldsymbol{r}(\theta,t), \{\boldsymbol{r}'_i(t)\}, t)|^2 \mathrm{d}\theta}{\int_\mathcal{B} |I(\boldsymbol{r}(\theta,t), \{\boldsymbol{r}'_i(t)\}, t)|^2 \mathrm{d}\theta} \tag{4.12}$$

where $\mathcal{B}$ is the boundary of the particle. Since the particle has a fixed radius a we parametrise the different positions at the boundary by using the angle θ. By using $\langle \nabla I \rangle_\circ$ we introduce the weighted averaged mentioned above.

A key step in our simulations is determining the initial positions of the particles. At the start of the simulations the particles need to be in a state comparable to dim-light adapted chloroplasts, i.e. the particles have a high packing fraction and are spread out over the cell in a disordered fashion (see first panel Fig. 11c). The packing fraction ϕ of the system needs to be taken into account in order to achieve this. We know that dim-light adapted chloroplasts have a packing fraction of approximately 0.71 [33]. Consequently, we want our system to have a fixed packing fraction, independently of the number of particles. This is done by making the boundaries of the cell dependent on the packing fraction and the number of particles:

$$x_{\text{max}} = \frac{\sqrt{N \pi a^2}}{4n\phi} \tag{4.13}$$

$$y_{\text{max}} = n x_{\text{max}} \tag{4.14}$$

with

$$\phi = \frac{N A_{\text{particle}}}{A_{\text{cell}}} \tag{4.15}$$

Here x_{max} and y_{max} are the upper boundaries of the cell, N is the number of particles in our system and n is the ratio between the height and width of the cell. A_{particle} and A_{cell} are the areas of the particle and cell respectively. We define the lower boundaries of the cell as $x_{\text{min}} = -x_{\text{max}}$ and $y_{\text{min}} = -y_{\text{max}}$, such that $A_{\text{cell}} = 4x_{\text{max}}y_{\text{max}}$. We set $n = \frac{1}{2}$ to get a rectangularly shaped cell, similar to the shape of an *Elodea densa*

Parameter	Definition	Values used in simulations
(x_0, y_0)	Initial position of a particle	No fixed values
a	Radius of the particle	1
ϕ	Packing fraction	$[0.04, 0.71]$
D	Diffusion constant	0.1
β	Coupling to the light field	-1
dt	Integration time step	10^{-4}
t_{tot}	Total integration time	10^5
A	Amplitude of the external light field	$[5, 200]$
C	Amplitude of the shadow	A
σ	Distance at which U_{cs} drops of to zero	$0.5a$
ϵ	Depth of the potential well of U_{cs}	1
r_0	Center of the Gaussian of U_{cs}	0.3 or 0.5
c_0	Width of the Gaussian of U_{cs}	1.3 or 1.5
u_0	Depth of the Gaussian of U_{cs}	5 or 25

Table 2: Simulation parameters of the agent-based Brownian dynamics model for chloroplast motion.

cell [33]. The initial position of each particle is determined by placing all particles at random positions inside the cell at a minimum distance a (particle radius) from each other. Even though this algorithm gives us the desired state, this procedure is extremely slow for system with more than 4 particles, so we use a different initialisation method for many particle systems.

Instead, we start by determining the length of the side of a square unit cell with the correct packing fraction ϕ using the following equation,

$$l_{\text{unit cell}} = \frac{\pi a^2}{\phi}. \tag{4.16}$$

Next we calculate the amount of unit cells needed in both the vertical and horizontal direction to fit all N particles. This is done by factorising the total number of particles into primes and multiplying the first half of these factors to get the height of the cell and the second half to get the length of the cell. With this method we make a rectangularly shaped cell filled with N particles in a grid. After the particles have been placed in a grid we run a simulation without an external light potential with a total length of 10^4 time steps and we decrease the diffusion coefficient gradually from $D = 0.5$ to $D = 0.1$ during $2/3$ of the time total time and let the system equilibrate for $1/3$ of the total time. This gives us a system of in a state similar to dim-light adapted chloroplasts as the initial condition of our simulation, see Fig. 16a.

All model parameters and their values in our simulations are shown in the Table 2. All parameters in the simulations have arbitrary units.

4.4   Numerical simulations

4.4.1   Model verification

In order to test the response of the particles to a light gradient we run a simulation where half of the cell is in the light and the other half is in the shadow. For this simulation we use Eq. (4.6) with $t_{\text{on}} = 0$ and Eq. (4.5) with $\alpha = 10$ to create a constant light field on the right side of the cell and an overall shadow on the left side. Fig. 13 shows snapshots of the particles at different times of the simulation. At $t = 0$ half of the purple particle is illuminated, while the red particle is fully illuminated. The purple particle is subjected to a light gradient and rapidly moves away from the light. After it has moved into the shadow region it stays there. The red particle is not subjected to a light gradient and consequently does not move out of the light. It eventually moves to the shadow, but only after diffusively moving close to the boundary between shadow and light. This type of motion is qualitatively comparable to the dynamics of chloroplasts seen in the experiments performed by Kong *et al.* [27], see Section 4.1. This shows that our model captures the basic response to a light stimulus well.

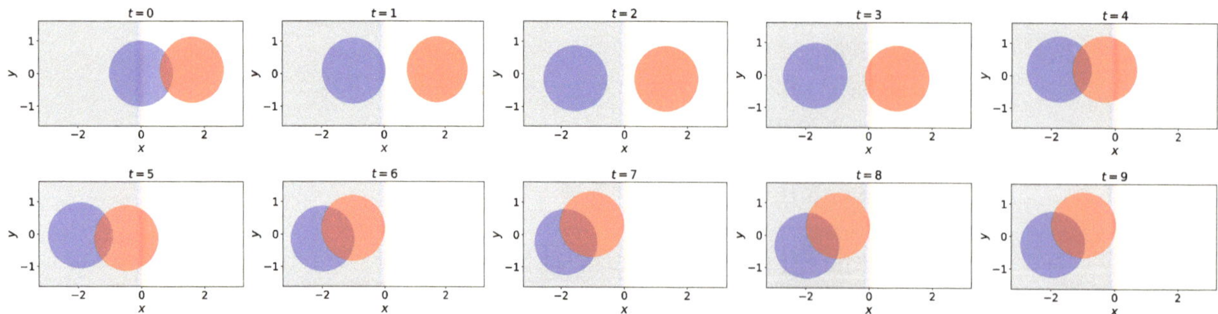

Fig. 13: Snapshots of the motion of two model particles inside a cell. The cell is shaded for $x < 0$ and illuminated for $x > 0$. The purple particle feels a light gradient at $t = 0$, while the second red particle does not. The purple particle quickly moves into the shadow, while the red one only does so after having diffusively moved close to the shaded region. In all figures t indicates the simulation time.

For the simulation described here we use the following parameters: $\phi = 0.3$, $A = 5$, $\sigma = 5$, $r_0 = 0.3$ and

$c_0 = 1.3$. With this set of parameters we observe particle overlap when the light intensity is zero. In later iterations of this model we will use a stronger repulsive potential.

4.4.2   Comparison with experiments

Having developed and tested the model, we compare it to experimental measurements. Since the measurements in experiments are done on single chloroplasts, we run multiple single particle simulations for these comparisons. Given that we are simulating a single particle system, there is no interaction potential or shadow effect between particles, i.e. $U_{cs}\left(\boldsymbol{r}(t), \{\boldsymbol{r}_i'(t)\}, t\right) = 0$ and $I_{\text{shadow}} = 0$, for these simulations. In experiments the chloroplasts are illuminated by a light beam with a certain width that is turned on after a set amount of time. We simulate a similar system by using Eq. (4.17) to create a light beam centered at at the origin, see Fig 14a. We use Eq. (4.6) to turn the light on at a specific time t_{on}.

$$S(x, y, t) = \begin{cases} 1 & \text{if } x \leq w \text{ and } y \leq h, \\ 0 & \text{if } x \geq w \text{ or } y \geq h, \end{cases} \tag{4.17}$$

with w the width of the shadow and h the height of the shadow. For all simulations described below we set the following parameters: $\phi = 0.04$, $t_{\text{on}} = \frac{t_{\text{tot}}}{2}$ and $w = h = 2a$. We perform a parametric sweep over multiple light intensities in the range $A = [20, 200]$ with $\Delta A = 30$. For every configuration we perform 50 runs.

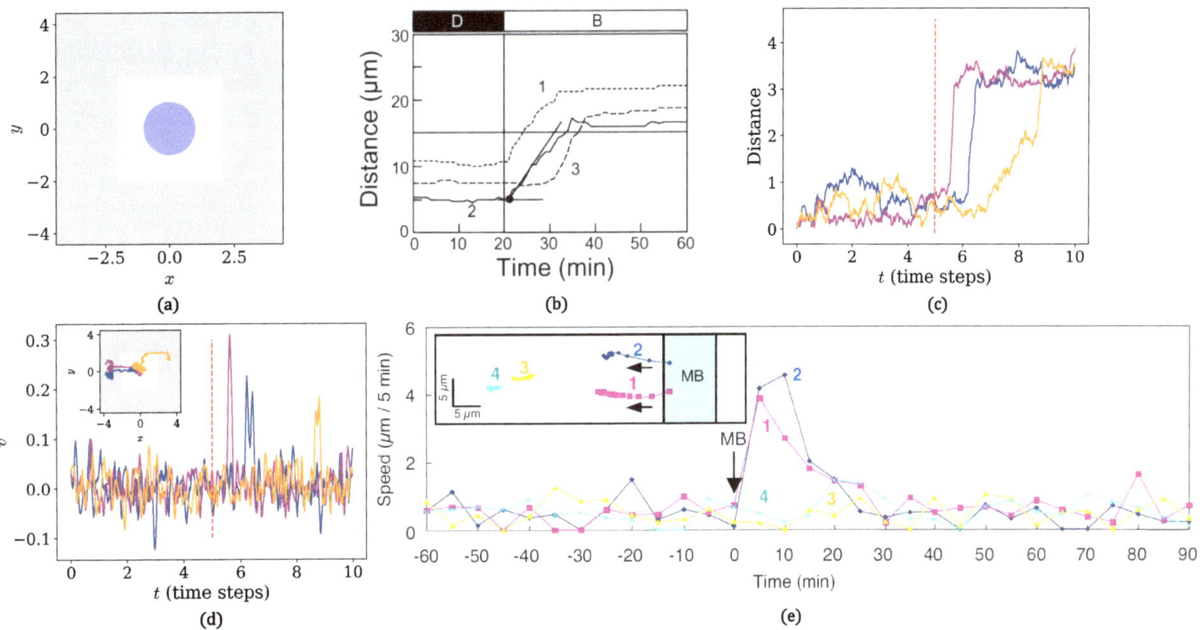

Fig. 14: **(a)** System configuration for our simulations. A single particle is placed inside a square light beam that is turned on at t_{on}. **(b)** Distance between three chloroplasts and the center of a microbeam measured by Kagawa *et al.* [69]. Each line represents a different chloroplast. D: dark incubation (light off), B: blue light irradiation (light on). Reprinted from [69] with permission from Royal Society of Chemistry. **(c)** Simulation results for the distance between the center of a light beam with intensity $A = 80$ and the center of mass of three particles. Each line represents a different particle. Red dashed line: indicates t_{on}. **(d)** Velocity of the simulated particles as a function of time. Red dashed line: indicates t_{on}. (Inset): trajectories of the center of mass of each particle. Light field indicated in the background. **(e)** Velocity of chloroplasts measured by Kadota *et al.* [20]. Control chloroplasts outside the irradiated area are indicated by cyan and yellow. Chloroplasts in the irradiated area are indicated by blue and magenta. (Inset): chloroplast positions at 5-min intervals after irradiation. Reprinted from [20] with permission from PNAS.

First, we look at the distance between the center of the light beam and the particle over time. Kagawa *et al.* [69] measured this for 3 chloroplasts irradiated by a high-fluence-rate blue light microbeam. The results of their measurements can be seen in Fig. 14b. After the chloroplasts are irradiated, the distance between the microbeam and the particles increases rapidly until the chloroplasts is outside of the light. After this the distance plateaus again and remains constant. Our model particles shows very similar behaviour, see Fig. 14c. The results of our simulations are much noisier than the experimental measurement since the particles undergo Brownian motion, which is not measured in experiments.

Furthermore, Kadota *et al.* [20] measured the velocity of chloroplasts during the dim-to-bright light transition, see Fig. 14e. We see that the irradiated chloroplasts experience a sharp increase in speed while they are moving away from the light. This is in agreement with the measurements by Kagawa *et al*; the jump-like motion of the chloroplasts indicates a sharp increase in velocity while the particles move away from the light. When calculating the velocity of the model particles, see Fig. 14d, we find that there is indeed a peak whenever the particles move away from the light. These results show that the single particle dynamics of our model are comparable to experimental measurements. This indicates that this simple model captures the key features of single chloroplast motion during the dim-to-bright light transition.

Lastly, Kagawa *et al.* calculated the velocity of the chloroplasts as a function of the light intensity, see Fig. 15a. They calculated the velocity as "the slope of the distance travelled during the avoidance movement and the distance between the starting of light irradiation and the starting of chloroplast avoidance movement" [69]. We do a similar analysis of the velocity of the particles for different light intensities, see Fig. 15b. Both the experiments and simulations show an increase of the velocity with light intensity, however the dependency is not the same. Kagawa *et al.* state that the experimentally measured velocity for the wild type (WT) chloroplasts increases linearly with logarithmically increasing light intensity [69]. In our simulations, the increase is more than linear when plotting the intensity in a logartihmic scale. We do see a linear increase with respect to linearly increasing intensity (see inset in Fig. 15b). This is what we expect from our equations of motion. For this system we have:

$$\nabla I_{\text{total}}(\boldsymbol{r}(t)) = A\delta(\boldsymbol{r}(t) - \boldsymbol{r}_b) \tag{4.18}$$

with $\boldsymbol{r}_b$ the position of the boundary of the light beam (see Fig. 14a). Plugging this into Eq. (4.2) we see that in our model the velocity should indeed increase linearly with the light intensity A. The difference between experiments and simulations can be attributed to a number of different factors. First and foremost, there are key differences between the system of chloroplasts and the model system. Additionally, the data from experiments is quite sparse which makes it difficult to make a thorough comparison with simulation results. All of these factors will be discussed in depth in Section 5. Nonetheless, all our results are qualitatively comparable to the experimental results, which shows that our model captures the dynamics of single chloroplast motion. This leads us to conclude that light gradients do indeed drive the motion of chloroplasts.

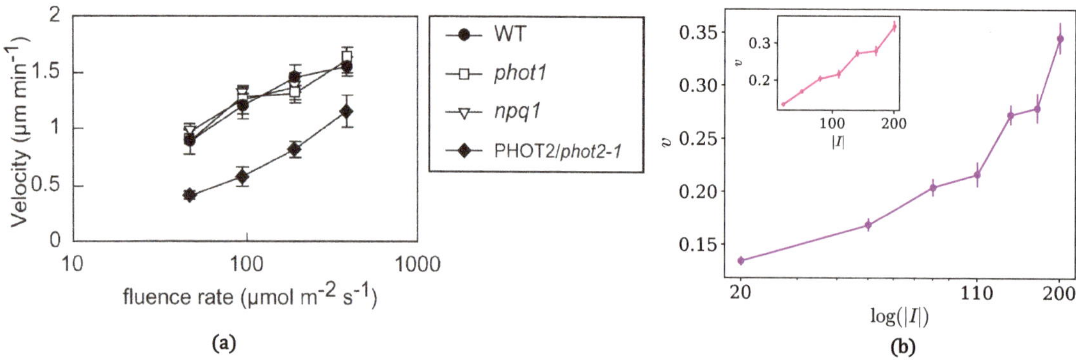

Fig. 15: (a) Velocity of chloroplasts for different light intensities as measured by Kagawa *et al.* [69]. The different data points are for different plant mutants observed in experiments. We concern ourselves only with the results for the wild type (WT). Reprinted from [69] with permission from Royal Society of Chemistry. (b) Velocity of the model particles moving away from the light as function of the logarithmic light intensity. (Inset): Velocity as a function of light intensity on a linear scale.

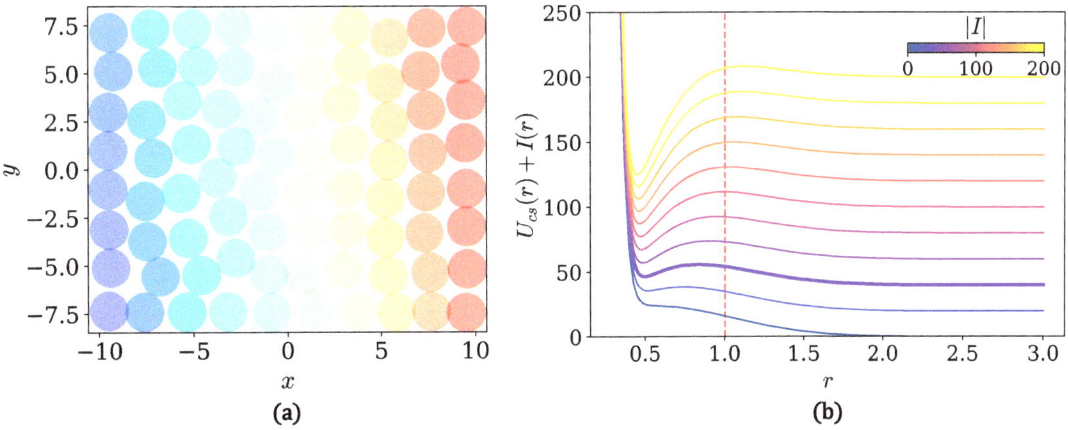

Fig. 16: **(a)** Initial position of the 80 particle system. This state was reached by gradually decreasing the diffusion coefficient from $D = 0.5$ to $D = 0.1$ during 10^4 times steps. **(b)** Effective potential the system is subjected to for different light intensities with $\Delta|I| = 20$. The potential is almost entirely repulsive for $I < 40$ (thick line). For larger intensities we see a global minimum appear in the potential that increases in depth with intensity.

4.4.3 An eighty-particle system

Having tested the dynamics of our model on a single particle level, we move on to simulations of a many particle system. We do simulations with 80 particles, which is comparable to the number of chloroplasts inside an *Elodea densa* cell. The starting position of the 80 particle system can be seen in Fig. 16a. This system has a similar packing fraction to chloroplasts in the dim light adapted state (≈ 0.7) [33]. The potential between the particles has been chosen in such a way that the particles can only overlap if the external potential, i.e. the light intensity, is strong enough. We run simulations with a constant external light field, so we set $S(x, y, t) = 1$ and use Eq. (4.6) with $t_{\text{on}} = 0$. We perform a parametric sweep over eight different light intensities A in the range $[20, 200]$ with $\Delta A = 30$. This range of intensities was chosen based on the shape of the effective potential at different light intensities, see Fig. 16b. For all simulations discussed below we set the following parameters: $\phi = 0.71$, $\sigma = 0.5$, $r_0 = 0.5$, $c_0 = 1.5$ and $u_0 = 25$.

We start our analysis of the system dynamics by studying the positions of particles and the nearest neighbour (NN) connections at the final time of our simulations, see Fig. 17. Particles are defined to be nearest neighbours if the distance between them $\leq 2a$ with a the particle radius. We can conclude that the number of NN of each particle increases as the light intensity increases, which indicates that the particles cluster more with increasing intensity. This shows that the effective attraction between the particles due to the external light field leads to clustering. However, the clustering dynamics in our simulations are different from what we see in chloroplasts. The clusters in our system are always static, while chloroplast clusters in *Elodea densa* can move inside the cell. This suggests that, while the effective attraction between the particles does lead to clustering, it does not induce the dynamics we see in experiments. Consequently, other physical effects need to be added.

Nonetheless, we investigate the clustering dynamics as a function of the light intensity to gain a better understanding of our current model. Firstly, we calculate the probability density distribution (PDF) of inter-particle distances over time for different light intensities, see Fig. 18a-g. For $I < 80$ we see clear bands in the pdf, which means we have a long-ranged periodic structure and there is no cluster formation. This is indeed what we see in Fig. 17a and b. For larger intensities these bands start to disappear and we see one large region of high probability. Additionally, the mean inter-particle distance (Fig. 18h) decreases as the intensity increases. This implies there is cluster formation, which can indeed be seen in Fig 17c-g. The change in the mean inter-particle distance gives us an indication of the clustering speed, see Fig. 18i. We see that the particles cluster much faster as the intensity increases. With more experimental measurements we could compare these values to measurements on chloroplasts to see if our model behaves comparably.

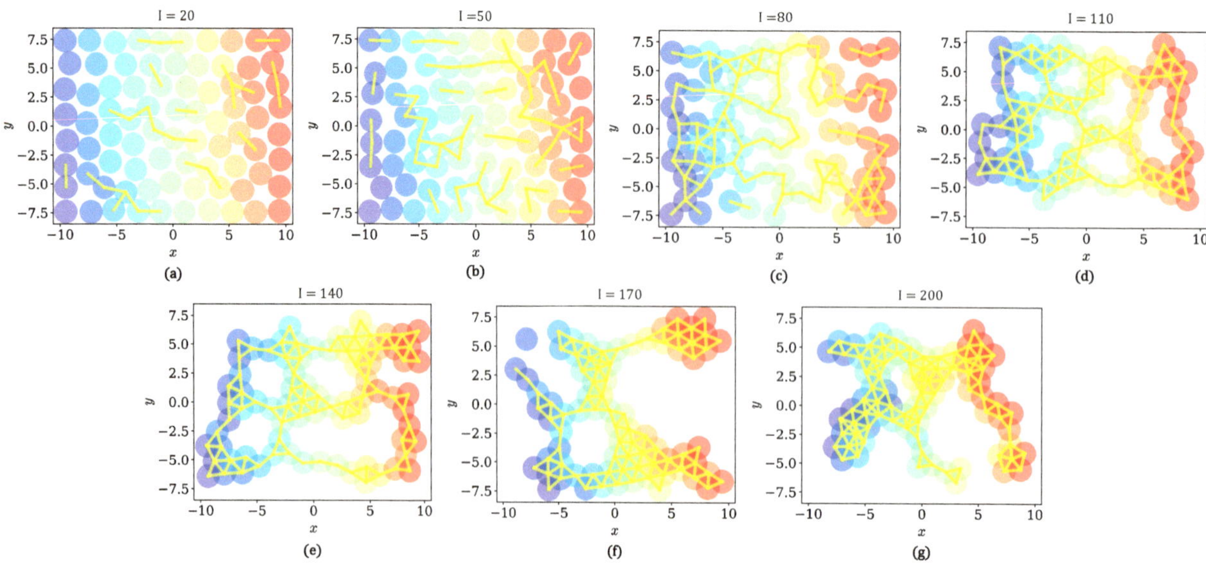

Fig. 17: Final position of an 80 particle system for different light intensities I. The yellow lines indicate the nearest neighbours (NN). We define particles to be NN if the distance between them $\leq 2a$, with a the particle radius.

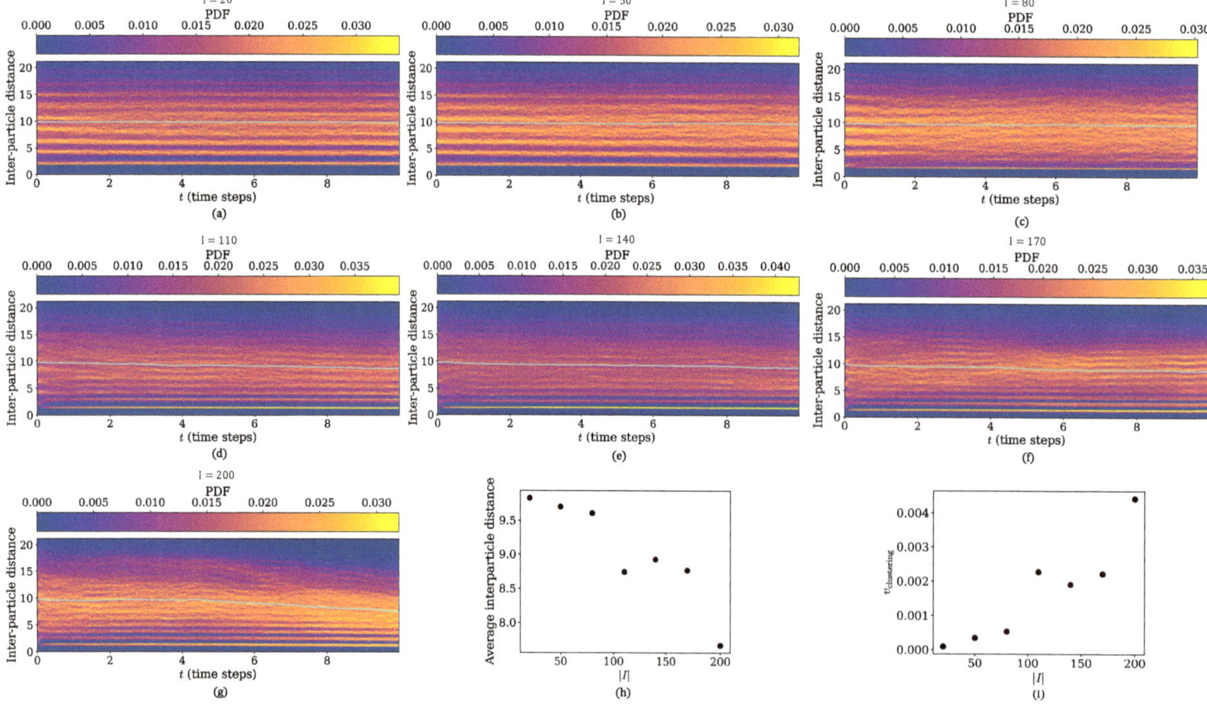

Fig. 18: (a - g) Probability distribution function (PDF) of the inter-particle distance for different intensities I. Bands in the PDF indicate the existence of a long-ranged periodic structure. Blue line: average inter-particle distance. (h) Average inter-particle distance for increasing light intensities. (i) Clustering speed as a function of light intensity. The clustering speed was calculated by taking the average of the derivative of the mean inter-particle distance.

5 Discussion

5.1 The dim-light toy model

The results from our Langevin model (Eq. (3.4)) for the accumulation response are in good agreement with experimental measurements. This indicates that the burst-like re-arrangements of chloroplasts might be triggered upon exceeding a threshold similar to the cage length, the space each particle is permitted before encountering its neighboring particles, in glasses [70]. However, our model is not unique in describing the dynamics of a system close to the glass transition. There are other models that have similar statistics, for example the continuous time random walk (CTRW) model proposed by Chaudhuri et al [38, 39]. This model is used to describe a universal step-size distribution function in various systems close to the glass transition. However, the fit parameters are different for both models. The CTRW model needs the mean waiting time between jumps as a fit parameter, but this is a difficult parameter to infer from limited experimental data. Our toy model enables us to convert the waiting time between jumps into a length scale x_c. This length scale is a parameter that is more easily extracted from experimental measurements. Our model could also be applied to study the stability of other systems close to the glass transition. The simplicity of our model is one of its strong points, since it is easy to understand and has only a few parameters. However, this simplicity limits the applications of the model for further research into the dynamics of chloroplasts. This model can only be used to study the dynamics in the dim-light adapted state; more elaborate schemes are needed to describe the dynamics of chloroplasts in other light intensity regimes

The simplicity of our toy model also leads to differences between the results from experiments and simulations. Chloroplasts are anchored to the plasma membrane [66] which allows for mostly two-dimensional motion. Consequently, the system of chloroplasts in the dim-light adapted state can be approximated as a tightly packed quasi-2D layer. This system shows the hallmarks of dynamical heterogeneity, the coexistence of fast and slow particles. The toy model is a one-dimensional single particle system, therefore there can be no coexistence of different particles. The dynamics of the many particle system are simplified by using the spring to model the confinement due to slow-moving neighbours and the 'jump element' to model the burst-like rearrangements. We perform multiple runs of a simulation in order to generate statistics for the system, but this does not approximate a many particle system with true dynamical heterogeneity. In experiments we see that the trajectories of multiple particles can be correlated during the sudden re-arrangements, since neighbouring particles interact inside the plant cell (Fig. 6a). Evidently, a one particle model cannot show such correlations. Furthermore, there is a difference in the scaling of the MSD on large time scales. In the experiments (Fig. 6b) all regimes approach a diffusive scaling on large time scales. In the simulations (Fig. 9) the scaling of the MSD goes from diffusive to subdiffusive on large time scales as x_c or D increase. This difference is due to the fact that, on large time scales, the model particles are caged for high x_c or D, while the chloroplasts are essentially freely diffusing[8]. The version of the model with diffusive diffusivity also neglects the sub-diffusive dynamics that we see in experiments by construction [53, 54]. Despite the simplifications, this model captures the dynamics of dim-light adapted chloroplasts qualitatively and quantitatively. The model is sufficient to predict that, not only the activity σ, but also the critical size x_c can easily drive the system out of the caged dynamics.

5.2 The multi-particle Langevin model

The agent-based Langevin-model (Eq. (4.2)) captures the basic dynamics of single chloroplasts motion qualitatively, see Fig. 14 and 15. This shows that the hypothesis, that chloroplast motion is driven by light gradients, is correct. However, even though experiments and simulations both show and increase of the velocity with the light intensity, the trend of the increase is different. This could be due to a number of different factors. Firstly, there are some key differences between the biological system and our model system. Chloroplast velocity saturates at a certain light intensity, since there is a finite amount of actin binding sites on the chloroplast outer envelope. Also, we have learned from experimental observations that chloroplasts can 'die' under very high light intensities because this irreparably damages the photosynthetic apparatus inside them [71]. At this point the chloroplast will stop moving completely. Both of these effects are not taken into account in our model and this will affect the shape of the velocity curve. These effects should be

---

[8]Because the chloroplasts move so slowly that they do not touch other particles around them.

added to the model in future work, for example by using an equation inspired by Michaelis-Menten kinetics. This equation models the rate of product formation as a function of the concentration of a limited substrate. Using this framework one could find an equation for the velocity of a chloroplast that depends on the concentration of available actin binding cites, i.e. activated CHUP1. Since the number of active binding sites depends on the intensity of light, one could find a saturation intensity and use this to couple the velocity of the particle to the light intensity. Furthermore, the comparison between experimental and model results is not very robust since Fig. 15a only has four data points. More experimental data is needed to be able to make a more thorough comparison.

Unlike the single particle case, our model does not capture the full dynamics of a many chloroplast system. Experiments show that chloroplasts rapidly form tightly packed clusters after the a high-intensity light source is turned on (Fig. 11c). These clusters of chloroplasts can move inside the cell. Even though the particles in our model do form clusters as the light intensity increases, these clusters never move. Moreover, the clusters seen in experiments are circular, while in simulations the clusters form a network. From this we conclude that the effective attraction between the particles due to the light does lead to clustering. However, other physical effects need to be added in order to reproduce the collective dynamics seen in experiments. Consequently, the model needs to be revised and other physical effects need to be added.

First and foremost, chloroplasts do not only sense light locally, but they are capable of long-range sensing since they can receive signals from photoreceptors located on the plasma membrane (see Section 2). This is not taken into account in the current version of the model and might lead to long-range interaction between the particles. Secondly, in experiments the plant cell is not irradiated by a constant homogeneous light field, but the light source can fluctuate. The introduction of such heterogeneous light field into the model could have a large impact on the particle dynamics, since the particles are very sensitive to changes in the light field. This might lead to the type of clustering seen in *Elodea densa* cells during experiments. Another possible addition are non-reciprocal interactions between the particles. Chloroplast can overlap in the z-direction, i.e. a chloroplast can be on top of another one. Consequently, the chloroplast on top gives shade to the one below, but the lower chloroplast does not give shade to the upper one. Adding this to the model might lead to a 'chasing' effect where the lower particles feel an effective attraction to the upper particles, since those provide shade, but the upper particles move away from the lower particles since they do not feel any attraction but only repulsion[9]. It is possible that this effect could lead to the motion of the clusters that we see in experiments. Furthermore, the current version of the model does not take into account the existence of torque between the particles. Even though single chloroplasts do not experience torque, as was explained in Section 2, there is torque due to friction between the chloroplasts. This might lead to particle rotation and could have an overall effect on the clustering dynamics. The bi-layered nature of *Elodea densa* might also affect the dynamics. The top and bottom layer of cells can interact with each other and the motion of the chloroplasts in both layers might be correlated. In order to understand the interactions between the layers more experimental measurements are needed. Lastly, hydrodynamic interactions could be added since chloroplasts move inside the cytoplasm, which is an active viscoelastic medium.

There might be another reason for the differences in clustering dynamics between experiments and the model. Chloroplasts are active particles. Consequently, they continuously consume energy to be able to self-propel, which means they are intrinsically out of thermal equilibrium. As a result of this, active systems do not obey time-reversal symmetry and therefore their steady state does not obey the principle of detailed balance[10]. Because of this, Motility-Induced Phase Separation (MIPS) can take place. This is the phenomenon where an active system phase separates due to the non-equilibrium character of the single-particle motion. MIPS occurs because active particles tend to accumulate in regions where they move slowly since they spend more time in low velocity regions, which increases the particle density. This increase in particle density decreases the velocity of the particles even more. This creates a positive feedback loop that leads to phase separation [63, 64]. From experiments we know that chloroplasts have different velocities depending on the intensity of light and we know that the particles tend to accumulate in regions of low-intensity light, where they have a small velocity. Consequently, the clustering effect we see during the dim-to-bright light transition could be due

---

[9]This repulsion comes from the quasi 3D description in 2D that is used in our model. As mentioned in Section 3.2, the repulsive potential models the energetic cost of particles overlapping in the z-direction.

[10]This principle states that, at equilibrium, every process is in equilibrium with its reverse.

to MIPS. Phase separation and self-assembly in light sensing particles has been studied in multiple systems from bacteria to active colloidal particles [62, 72, 73].

Contrary to chloroplasts, our model particles are not self-propelled. They are passive particles responding to an external gradient, so the system is intrinsically in thermal equilibrium. Such a system cannot undergo MIPS, since this is a phenomenon inherent to active systems. Introducing activity of the particles into the current Langevin model in order to study MIPS should be a focus point for future iterations of the model. This can be done by using an Active Brownian Particle (ABP) model [74] or an Active Ornstein-Uhlenbeck Particle (AOUP) model [75, 76]. The ABP introduces a constant self-propulsion term to the overdampend Langevin equation. Consequently, this model requires a separate description of the orientational dynamics of the particle, which is done by introducing diffusional relaxation of the swimming direction. The AOUP model replaces the self-propulsion term in the ABP model with a translational noise with memory (coloured noise), and therefore does not require a separate description of the orientational dynamics. Either model could be used, since they are both computationally straightforward. Nonetheless, finding a correct description of the orientational dynamics of the particle might be difficult, which makes the ABP model less appealing. The AOUP model seems like a better fit to update our current model as experimental measurements could be used to determine the type of noise present in the system of chloroplasts. By combining such experimental measurements with the AOUP model, an accurate description of the particle dynamics could be achieved. Additionally, the dimensionality of the system can influence the phase behaviour of active brownian particles [77]. Consequently, the effects of dimensionality on the model system should be investigated thoroughly in future studies, to determine if a 2D model is enough to model the dynamics of the 3D chloroplast system.

For this study, the model parameters were not determined through comparison with experimental measurements due to a lack of experimental data. The parameters of the interaction potential $U_{cs}\left(\boldsymbol{r}(t), \{\boldsymbol{r}_i'(t)\}, t\right)$ were chosen according to the necessary shape of the potential. The values of the light intensity were chosen according to the shape of the effective potential at different light intensities. The parameter value of the diffusion constant was chosen through a qualitative comparison between the dynamics of model and experimental data. Finding better parameter values for the model could have an impact on the dynamics of the model particles and might lead to different results. Future work could focus on parameter estimation from experimental data, whenever more data is available. In order to make parameter estimation possible, the number of variables in the model should be reduced. This can be done by finding dimensionless parameters through a mathematical analysis of the equation of motion of the system. In future experiments one could quantify the overlap between chloroplasts inside a cluster relative to their size and use this as a measure to estimate the parameters of the potential. The amplitude of the shadow cast by the chloroplasts could be determined by measuring the intensity of light under a chloroplast whenever light shines on it. The difference in intensities between the external light source and the bottom of the chloroplasts could give a more accurate relation between the amplitude of the external field and the amplitude of the shadow.

From a computational point of view there are some points of improvement. Firstly, the current simulations are computationally expensive. A single simulation of an 80 particle system with a total of 10^5 time steps takes approximately five hours. This made it challenging to generate large amount of data, which consequently reduced the accuracy of the results for the many particle simulations. The numerical simulations should be made more efficient by using a different or updated solver to integrate the equations in future iterations of the model. Secondly, the integration time step is very small at the moment, see Table 2. If this parameter is increased, the simulation can become unstable. This is due to the fact that the particles in the simulation can overlap and the interaction potential between the particles is highly repulsive close to the center of the particle. If dt is increased, the particles can come too close to each other and the simulation blows up. This issue could be solved by using a different and more stable integration method. Lastly, for this study we used the `sdeint` package to integrate the Langevin equation, which works well, but has its limitations. There is not a lot of flexibility in the way in which the equations and initial conditions need to be structured to fit into the integration algorithm. This is a limiting factor if more complexity is added to the model. A new solver that is more flexible is needed to make the integration of a more complex version of the model possible.

5.3   One model to rule them all

Currently, there are two different models for the motion of chloroplasts under different light intensities, however, in future, one could make a single model for chloroplasts motion with light intensity as the driving parameter. The agent-based Langevin model proposed in this work contains all the building blocks needed to make such a model. A key step to achieving this is the parameter β, the coupling to the light field. This parameter needs to be dependent on the light intensity, since this drives the transition between repulsion and attraction to the light. Further experimental measurements are also needed. The transition between the accumulation and avoidance response needs to be observed, as well as the transition from collective clustering dynamics to the motion during the avoidance response. Additionally, the impact of confinement due to the cell walls on the transitions between dark and light-adapted states remains a matter for further investigation. A single unified model for chloroplast motion could lead to a phase diagram for chloroplasts and will add to our understanding of phase transitions in dense active biological matter, which is a current topic in active matter research [78].

6   Conclusion

Summarising, for this thesis we made multiple numerical models to study the dynamics of chloroplasts in a dim-light adapted state and during the dim-to-bright light transition. With these models we aimed to gain an understanding of the different types of dynamics seen in experiments and their underlying physics. Chloroplasts accumulate into a single layer in response to weak light in order to achieve a high surface coverage and therefore maximize photosynthetic efficiency [7]. Since we know that chloroplasts avoid high-intensity light [8], they must maintain the ability to undertake this avoidance motion efficiently without being hindered by other chloroplasts. Therefore, the ability to increase the activity and a sufficiently loose packing of chloroplasts is necessary. Our study shows that chloroplasts in dim-light conditions resemble a system close to the glass transition. Chloroplasts cluster inside the plant cell in response to short exposure to high-intensity light in order to minimize photodamage. Our model for the dim-to-bright light transition shows that light gradients drive chloroplast motion and clustering between particles, but other physical effects are needed to fully understand the collective dynamics of the chloroplasts seen in experiments. Further research is needed to determine which physical effects underlie these dynamics. Nonetheless, our agent-based Langevin model provides a basis for future research into chloroplast dynamics and for achieving a unified model of chloroplast motion.

References

[1]   Stephen Hales. *Vegetable Staticks: Or, an Account of Some Statical Experiments on the Sap in Vegetables: Being an Essay Towards a Natural History of Vegetation. Also, a Specimen of an Attempt to Analyse the Air, by a Great Variety of Chymio-statical Experiments; which Were Read at Several Meetings Before the Royal Society....* Vol. 1. W. and J. Innys, at the West End of St. Paul's, 1727.

[2]   Neil A. Campbell et al. *Biology A Global Approach.* 11th ed. Pearson Education Limited, 2018. ISBN: 978-1-292-17043-5.

[3]   Gustav Senn et al. 'Gestalts-und Lageveranderung der Pflanzen-Chromatophoren'. In: (1908).

[4]   Masamitsu Wada. 'Chloroplast movement'. In: *Plant Science* 210 (2013), pp. 177–182.

[5]   Masahiro Kasahara et al. 'Chloroplast avoidance movement reduces photodamage in plants'. In: *Nature* 420.6917 (2002), pp. 829–832.

[6]   Olga Sztatelman et al. 'Photoprotective function of chloroplast avoidance movement: in vivo chlorophyll fluorescence study'. In: *Journal of plant physiology* 167.9 (2010), pp. 709–716.

[7]   Eiji Gotoh et al. 'Chloroplast accumulation response enhances leaf photosynthesis and plant biomass production'. In: *Plant Physiology* 178.3 (2018), pp. 1358–1369.

[8]   Masamitsu Wada and Sam-Geun Kong. 'Actin-mediated movement of chloroplasts'. In: *Journal of Cell Science* 131.2 (2018), jcs210310.

[9]   Josef Anton Böhm. *Beiträge zur näheren Kenntniss des Chlorophylls.* éditeur non identifié, 1857.

[10]   B Frank. 'Über lichtwärts sich bewegende Chlorophyllkörner'. In: *Bot. Ztg* 29 (1871), p. 209.

[11]   Hidenori Tsuboi and Masamitsu Wada. 'Chloroplasts can move in any direction to avoid strong light'. In: *Journal of plant research* 124.1 (2011), pp. 201–210.

[12]   Agnieszka Katarzyna Banaś et al. 'Blue light signalling in chloroplast movements'. In: *Journal of experimental botany* 63.4 (2012), pp. 1559–1574.

[13]   Takeshi Higa et al. 'Actin-dependent plastid movement is required for motive force generation in directional nuclear movement in plants'. In: *Proceedings of the National Academy of Sciences* 111.11 (2014), pp. 4327–4331.

[14]   Masamitsu Wada. 'Chloroplast and nuclear photorelocation movements'. In: *Proceedings of the Japan Academy, Series B* 92.9 (2016), pp. 387–411.

[15]   Rituparno Mandal et al. 'Extreme active matter at high densities'. In: *Nature communications* 11.1 (2020), pp. 1–8.

[16]   Yann-Edwin Keta, Robert L Jack and Ludovic Berthier. 'Disordered collective motion in dense assemblies of persistent particles'. In: *arXiv preprint arXiv:2201.04902* (2022).

[17]   W Haupt and R Scheuerlein. 'Chloroplast movement'. In: *Plant, Cell & Environment* 13.7 (1990), pp. 595–614.

[18]   Masamitsu Wada, Franz Grolig and Wolfgang Haupt. 'New trends in photobiology: light-oriented chloroplast positioning. Contribution to progress in photobiology'. In: *Journal of Photochemistry and Photobiology B: Biology* 17.1 (1993), pp. 3–25.

[19]   Shingo Takagi. 'Actin-based photo-orientation movement of chloroplasts in plant cells'. In: *Journal of Experimental Biology* 206.12 (2003), pp. 1963–1969.

[20]   Akeo Kadota et al. 'Short actin-based mechanism for light-directed chloroplast movement in Arabidopsis'. In: *Proceedings of the National Academy of Sciences* 106.31 (2009), pp. 13106–13111.

[21]   Sam-Geun Kong et al. 'Rapid severing and motility of chloroplast-actin filaments are required for the chloroplast avoidance response in Arabidopsis'. In: *The Plant Cell* 25.2 (2013), pp. 572–590.

[22]   Kazusato Oikawa et al. 'Chloroplast unusual positioning1 is essential for proper chloroplast positioning'. In: *The Plant Cell* 15.12 (2003), pp. 2805–2815.

[23]   Serena Schmidt von Braun and Enrico Schleiff. 'The chloroplast outer membrane protein CHUP1 interacts with actin and profilin'. In: *Planta* 227.5 (2008), pp. 1151–1159.

[24] Jose A Jarillo et al. 'Phototropin-related NPL1 controls chloroplast relocation induced by blue light'. In: *Nature* 410.6831 (2001), pp. 952–954.

[25] Takatoshi Kagawa et al. 'Arabidopsis NPL1: a phototropin homolog controlling the chloroplast highlight avoidance response'. In: *Science* 291.5511 (2001), pp. 2138–2141.

[26] Tatsuya Sakai et al. 'Arabidopsis nph1 and npl1: blue light receptors that mediate both phototropism and chloroplast relocation'. In: *Proceedings of the National Academy of Sciences* 98.12 (2001), pp. 6969–6974.

[27] Sam-Geun Kong et al. 'CHLOROPLAST UNUSUAL POSITIONING 1 is a new type of actin nucleation factor in plants'. In: *bioRxiv* (2020).

[28] Craig W Whippo et al. 'THRUMIN1 is a light-regulated actin-bundling protein involved in chloroplast motility'. In: *Current Biology* 21.1 (2011), pp. 59–64.

[29] Takatoshi Kagawa and Masamitsu Wada. 'Brief irradiation with red or blue light induces orientational movement of chloroplasts in dark-adapted prothallial cells of the fern Adiantum'. In: *Journal of Plant Research* 107.4 (1994), pp. 389–398.

[30] Hidenori Tsuboi and Masamitsu Wada. 'Chloroplasts move towards the nearest anticlinal walls under dark condition'. In: *Journal of plant research* 125.2 (2012), pp. 301–310.

[31] Sam-Geun Kong and Masamitsu Wada. 'Molecular basis of chloroplast photorelocation movement'. In: *Journal of plant research* 129.2 (2016), pp. 159–166.

[32] Sam-Geun Kong et al. 'Both phototropin 1 and 2 localize on the chloroplast outer membrane with distinct localization activity'. In: *Plant and Cell Physiology* 54.1 (2013), pp. 80–92.

[33] Nico Schramma, Cintia Perugachi Israëls and Maziyar Jalaal. 'Chloroplasts in plant cells show active glassy behavior under low light conditions'. In: *bioRxiv* (2022).

[34] Nicoletta Rascio et al. 'Photosynthetic strategies in leaves and stems of Egeria densa'. In: *Planta* 185.3 (1991), pp. 297–303.

[35] A Witztum. 'Transcellular chloroplast banding patterns in leaves of Elodea densa induced by light and DCMU'. In: *Annals of Botany* 42.182 (1978), pp. 1459–1462.

[36] Bo Wang et al. 'Anomalous yet brownian'. In: *Proceedings of the National Academy of Sciences* 106.36 (2009), pp. 15160–15164.

[37] Thomas J Lampo et al. 'Cytoplasmic RNA-protein particles exhibit non-Gaussian subdiffusive behavior'. In: *Biophysical journal* 112.3 (2017), pp. 532–542.

[38] Pinaki Chaudhuri, Ludovic Berthier and Walter Kob. 'Universal nature of particle displacements close to glass and jamming transitions'. In: *Physical review letters* 99.6 (2007), p. 060604.

[39] Pinaki Chaudhuri et al. 'A random walk description of the heterogeneous glassy dynamics of attracting colloids'. In: *Journal of Physics: Condensed Matter* 20.24 (2008), p. 244126.

[40] Elliott W Montroll and George H Weiss. 'Random walks on lattices. II'. In: *Journal of Mathematical Physics* 6.2 (1965), pp. 167–181.

[41] Christoffer Åberg and Bert Poolman. 'Glass-like characteristics of intracellular motion in human cells'. In: *Biophysical journal* 120.11 (2021), pp. 2355–2366.

[42] Thomas E Angelini et al. 'Glass-like dynamics of collective cell migration'. In: *Proceedings of the National Academy of Sciences* 108.12 (2011), pp. 4714–4719.

[43] Bradley R Parry et al. 'The bacterial cytoplasm has glass-like properties and is fluidized by metabolic activity'. In: *Cell* 156.1-2 (2014), pp. 183–194.

[44] Willem K Kegel, van Blaaderen and Alfons. 'Direct observation of dynamical heterogeneities in colloidal hard-sphere suspensions'. In: *Science* 287.5451 (2000), pp. 290–293.

[45] Mark D Ediger. 'Spatially heterogeneous dynamics in supercooled liquids'. In: *Annual review of physical chemistry* 51.1 (2000), pp. 99–128.

[46] Liesbeth MC Janssen. 'Active glasses'. In: *Journal of Physics: Condensed Matter* 31.50 (2019), p. 503002.

[47] Paul Langevin. 'Sur la théorie du mouvement brownien'. In: *Compt. Rendus* 146 (1908), pp. 530–533.

[48] George Gabriel Stokes et al. 'On the effect of the internal friction of fluids on the motion of pendulums'. In: (1851).

[49] Eric R Weeks and DA Weitz. 'Properties of cage rearrangements observed near the colloidal glass transition'. In: *Physical review letters* 89.9 (2002), p. 095704.

[50] B Doliwa and Andreas Heuer. 'Cage effect, local anisotropies, and dynamic heterogeneities at the glass transition: A computer study of hard spheres'. In: *Physical review letters* 80.22 (1998), p. 4915.

[51] Matthew Aburn. *GitHub - mattja/sdeint: Numerical integration of Ito or Stratonovich SDEs*. 2022. URL: https://github.com/mattja/sdeint/blob/master/sdeint/integrate.py.

[52] Luigi M Ricciardi and Shunsuke Sato. 'First-passage-time density and moments of the Ornstein-Uhlenbeck process'. In: *Journal of Applied Probability* 25.1 (1988), pp. 43–57.

[53] Mykyta V Chubynsky and Gary W Slater. 'Diffusing diffusivity: a model for anomalous, yet Brownian, diffusion'. In: *Physical review letters* 113.9 (2014), p. 098302.

[54] Aleksei V Chechkin et al. 'Brownian yet non-Gaussian diffusion: from superstatistics to subordination of diffusing diffusivities'. In: *Physical Review X* 7.2 (2017), p. 021002.

[55] Toshihiro Toyota et al. 'Non-Gaussian athermal fluctuations in active gels'. In: *Soft Matter* 7.7 (2011), pp. 3234–3239.

[56] Benno Liebchen and Hartmut Löwen. 'Synthetic chemotaxis and collective behavior in active matter'. In: *Accounts of chemical research* 51.12 (2018), pp. 2982–2990.

[57] Yuping Chen and James E Ferrell. 'C. elegans colony formation as a condensation phenomenon'. In: *Nature communications* 12.1 (2021), pp. 1–10.

[58] Jean-François Joanny and Sriram Ramaswamy. 'Filaments band together'. In: *Nature* 467.7311 (2010), pp. 33–34.

[59] Andrea Cavagna et al. 'Scale-free correlations in starling flocks'. In: *Proceedings of the National Academy of Sciences* 107.26 (2010), pp. 11865–11870.

[60] Isaac Theurkauff et al. 'Dynamic clustering in active colloidal suspensions with chemical signaling'. In: *Physical review letters* 108.26 (2012), p. 268303.

[61] Ivo Buttinoni et al. 'Dynamical clustering and phase separation in suspensions of self-propelled colloidal particles'. In: *Physical review letters* 110.23 (2013), p. 238301.

[62] Jeremie Palacci et al. 'Living crystals of light-activated colloidal surfers'. In: *Science* 339.6122 (2013), pp. 936–940.

[63] Michael E Cates and Julien Tailleur. 'Motility-induced phase separation'. In: *Annu. Rev. Condens. Matter Phys.* 6.1 (2015), pp. 219–244.

[64] Joakim Stenhammar. 'An Introduction to Motility-Induced Phase Separation'. In: *arXiv preprint arXiv: 2112.05024* (2021).

[65] Marcel Meyer, Lutz Schimansky-Geier and Pawel Romanczuk. 'Active Brownian agents with concentration-dependent chemotactic sensitivity'. In: *Physical Review E* 89.2 (2014), p. 022711.

[66] Yuuki Sakai and Shingo Takagi. 'Roles of actin cytoskeleton for regulation of chloroplast anchoring'. In: *Plant Signaling & Behavior* 12.10 (2017), e1370163.

[67] TPO Nogueira et al. 'Tracer diffusion in crowded solutions of sticky polymers'. In: *Physical Review E* 102.3 (2020), p. 032618.

[68] Andreas Rößler. 'Runge–Kutta methods for the strong approximation of solutions of stochastic differential equations'. In: *SIAM Journal on Numerical Analysis* 48.3 (2010), pp. 922 952.

[69] Takatoshi Kagawa and Masamitsu Wada. 'Velocity of chloroplast avoidance movement is fluence rate dependent'. In: *Photochemical & Photobiological Sciences* 3.6 (2004), pp. 592–595.

[70] Vincent E Debets, Xander M de Wit and Liesbeth MC Janssen. 'Cage Length Controls the Nonmonotonic Dynamics of Active Glassy Matter'. In: *Physical Review Letters* 127.27 (2021), p. 278002.

[71] Rienk van Grondelle. 'The Structure of Biological Matter and Photosynthesis'. 2019.

[72] Jochen Arlt et al. 'Painting with light-powered bacteria'. In: *Nature communications* 9.1 (2018), pp. 1–7.

[73] Giacomo Frangipane et al. 'Dynamic density shaping of photokinetic E. coli'. In: *Elife* 7 (2018), e36608.

[74] Pawel Romanczuk et al. 'Active brownian particles'. In: *The European Physical Journal Special Topics* 202.1 (2012), pp. 1–162.

[75] David Martin et al. 'Statistical mechanics of active Ornstein-Uhlenbeck particles'. In: *Physical Review E* 103.3 (2021), p. 032607.

[76] Lennart Dabelow and Ralf Eichhorn. 'Irreversibility in active matter: General framework for active Ornstein-Uhlenbeck particles'. In: *Frontiers in Physics* (2021), p. 516.

[77] Joakim Stenhammar et al. 'Phase behaviour of active Brownian particles: the role of dimensionality'. In: *Soft matter* 10.10 (2014), pp. 1489–1499.

[78] Matteo Paoluzzi, Demian Levis and Ignacio Pagonabarraga. 'From motility-induced phase-separation to glassiness in dense active matter'. In: *Communications Physics* 5.1 (2022), pp. 1–10.

Chloroplast Dynamics: 10 Things Biologists Don't Want You To Know

Warning: The following piece of writing contains some jokes and a fair amount of sarcasm. Please continue with caution.

Plants are strange. They don't seem to do much of anything, and yet, almost every organism on earth is dependent on them for survival. Biologists have been studying plants ever since science became a thing and thanks to them we have learned a lot about these strange organisms. However, the documentation of their findings leaves much to be desired. Most of the time the figures they use in their articles are incomprehensible or they have so few data points that I have no idea how they ever managed to draw a conclusion. As a scientist, this irks me, because I am genuinely interested in biology. Not that physicists are much better mind you. I would say we are better at presenting our results neatly, but the jargon used in our articles is frightening, even to me. The biggest issue I have with physicists is that they seem to be predominantly interested in things like black holes, subatomic particles or quantum mechanics. They never bother to look at what is right in front of them, like the plant sitting on their windowsill. If they ever took the time to ask themselves "What is happening inside this plant?", they would see fascinating things. I am one of those few physicists that took an interest and decided to look into what happens inside plant cells whenever light shines on them. So let's talk plants.

Let's start at the beginning. Inside plant cells are tiny little biological machines called chloroplasts. What are chloroplasts? Chloroplasts are the powerhouse of the plant cell. They keep plants, and indirectly us, alive by performing photosynthesis. You know photosynthesis, the process of turning sunlight into sugar while expelling the oxygen we need to breathe. This fact alone makes chloroplasts interesting from a biological point of view. But, why would *a physicist* be interested in them? Well, that's because chloroplasts can move in very interesting ways depending on the light that shines on them and as a physicist I am contractually obliged to be interested in moving things. Let me explain. Chloroplasts, and consequently plant cells, are very sensitive to different light intensities. Too much light will damage the chloroplasts and too little leads to a shortage of energy for photosynthesis. Sadly, plants can't rapidly move towards or away from light whenever the light source around them changes. Luckily, biology is majestic and plants have evolved a mechanism to still be able to manage the amount of light the chloroplasts are exposed to. This mechanism involves, you guessed it, the movement of chloroplasts. When low-intensity light shines on a plant cell, the chloroplasts inside it move towards the brightest areas, the top and bottom of the cell, where they spread out in order to maximise the amount of light every chloroplast gets. This is called the **accumulation response**. If we go to the other extreme and shine very bright light on a plant cell, the chloroplasts move away from the brightest areas, towards the sides of the cell, to minimise their light absorption and prevent damage. This is called the **avoidance response**. In the pictures below you can see how chloroplasts are positioned inside a plant cell during the accumulation and avoidance response.

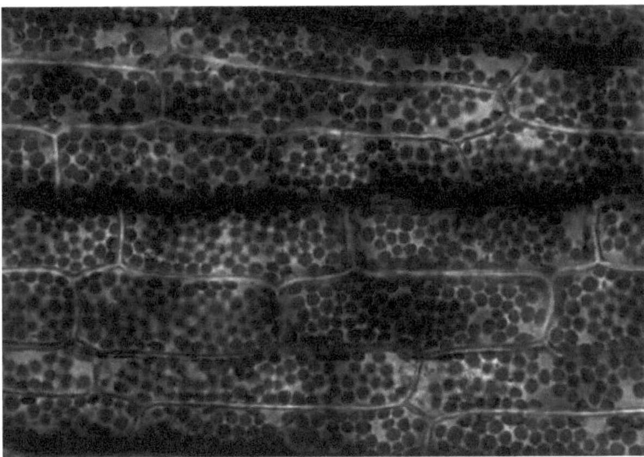

Accumulation response

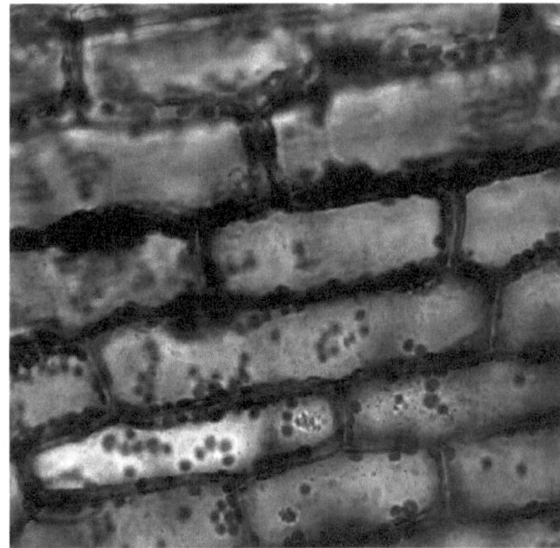

Avoidance response

Scientists have known that chloroplasts move like this since the second half of the 19th century. However, from a physics point of view, we don't understand how they can move in so many different ways. Actually, no physicist has ever bothered to look into this... until now. Together with my supervisors Mazi and Nico I went down this rabbit hole. There was one little thing though. Chloroplasts can move in so many different ways, we could never study everything. A year is much too short for that[11]. So, we decided I should focus on one specific type of motion that Nico was already looking into during his experiments: the **dim-light adapted state** of chloroplasts. This is fancy science language for 'chloroplasts that have been exposed to low-intensity light for some time'.

It all started with Nico doing some very fancy experiments where he filmed the chloroplasts in this particular state inside the cells of a water plant called *Elodea densa*[12]. He used these videos to track the motion of the chloroplasts and low and behold, we found something very interesting. The chloroplasts in the dim-light adapted state seem to be close to the glass transition. Now, I know that the term 'glass transition' sounds very strange, but don't panic, let me enlighten you. In physics a glass is not the object you drink water from, but it is a phase of matter just like a liquid, solid or gas. Glass is the phase in-between solid and liquid. Particles in a glass don't have a nicely arranged structure like a solid, but they also don't move around as much as particles in a liquid. A glass is essentially a very disordered and messy solid. Consequently, the glass transition is just the transition from a glass to a liquid. You might be thinking, how can chloroplasts or anything biology related be a glass, that sounds bizarre. And yeah, it is, but it's not uncommon. There are many biological systems that can be glass or that are close to the glass transition. For example, the cells of an embryo can go from a liquid to a glass state during the early stages of development[13]. If you are thinking, wow this is funky and interesting, please tell me more, you can read this article that I wrote all about the glass transition in biology. The article is in Dutch, so sorry to all non-Dutch speakers.

Anyhow, going back to the matter at hand. Since we found that chloroplasts in the dim-light adapted state seem to be close to the glass transition, we wanted to understand when they would be a glass and when they would be a liquid. Since this is very tricky to do experimentally, we came up with a simple mathematical model to do this, because I am still a theoretical physicist so math is required! I spent quite some time writing a working code for this model and running the simulations. I will not go into all the details of this, I don't think you're here for a math, physics and programming lecture. If you do want to know more, you can read Section 3.2 of my thesis. Using this model we found that the chloroplasts in the dim-light adapted state can go from a glass to a more liquid state depending on how close they are to other chloroplasts. Let me give you a visual example of how this works. Imagine you are at a concert and you are trying to walk from one side of the room to another. If the band is popular and the concert is sold out, there will be a lot of people and it will be very difficult for you to move around. You will constantly be bumping into everyone and progress will be slow. If, on the other hand, the band is not very popular and only a few people showed up, then you can easily move from one side of the room to another without bumping into anyone. This is essentially what the chloroplasts experience when they are in either a glass or liquid state. If neighbouring chloroplasts are very close to each other they will not have a lot of space to move around and they will be stuck (glass), while they will can move more freely (liquid) if the neighbouring particles are far away.

I thought this was already a good thesis-worth of knowledge I had acquired, but by this point I had only been working on this project for less than six moths and I had a whole year to fill. This meant we needed to come up with something else for me to do. Getting a master degree is hard work, can't be slacking off! Plus science is a lot of fun and I was so fascinated with chloroplasts at this point, I needed to know more. Since we had essentially only looked at the accumulation response, I suggested we look at the other side of the coin: the avoidance response. Nico had made some very fun videos where you could see how chloroplasts inside *Elodea densa* cells that started out in the dark would cluster after turning on high-intensity light. Essentially what happens is that chloroplast are napping in the dark and then we turn on a very bright light, which spooks them and they start forming clusters to get away from it. I decided to call this the dim-to-bright light transition. Here is a picture of what it looks like.

---

[11]Yes, science takes a long time. A year is nothing in science.

[12]I use the fancy biological name in order to impress you since this plant is called 'large-flowered waterweed' which sounds way less cool. It does come from South-America so that makes it a bit more cool.

[13]This is actually very important for correct embryonic development and a really fascinating area of research.

Dim to strong light transition

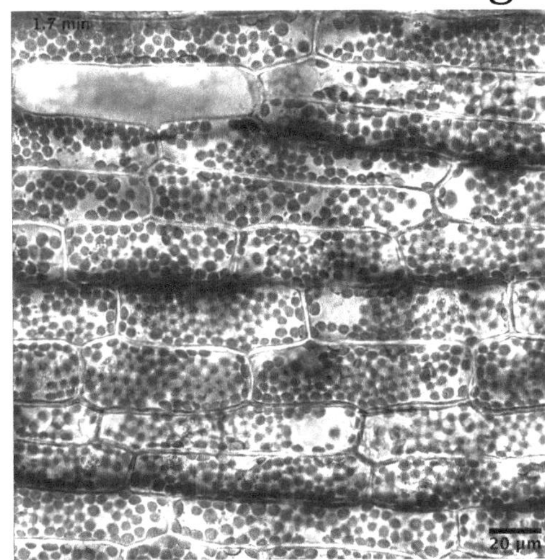 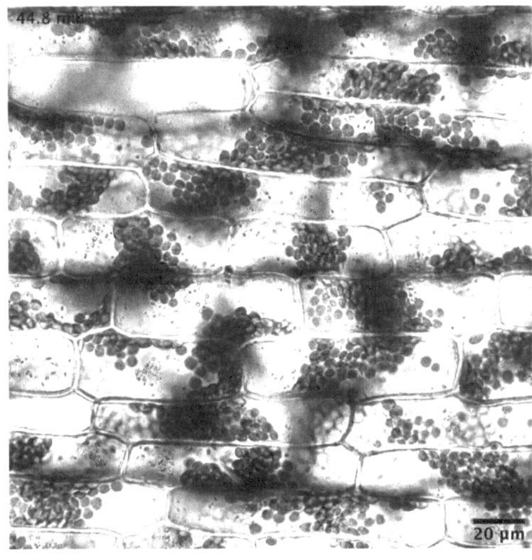

Chloroplasts in the dark   Chloroplasts after a bright
light is turned on

Inspired by the videos we decided to focus on studying the dim-to-bright light transition. This part of the project was a bit different to the what we had done previously. For the first part of the project it was mostly Mazi and Nico coming up with nice ways to model the chloroplasts based on the experimental measurements and I was mostly coding and doing data analysis. However, this time no one had a clear idea of how we were going to tackle the modelling aspect and Nico did not have time to do any of his fancy tracking experiments, because he was quite busy writing a super nice paper on our study of the dim-light adapted chloroplasts (here is the reference for that [33] in case you are interested). As a consequence, I was given free reign to develop a model, which was quite terrifying but also really cool and challenging. I started by doing a deep dive into the literature on chloroplast motion. After reading more papers than I care to count[14], I came up with an idea for how to model chloroplast motion during the dim-to-bright light transition. Again, I will not explain how the model works here, that's what the thesis is for. Everything is explained in Section 4.2 in excruciating detail.

Since we really did not know why chloroplasts move the way they do during the dim-to-bright light transition, we thought we could use this model to test some hypotheses we had. Firstly, from my literature deep dive we had learned that, whenever there is high-intensity light, chloroplasts that are partially in the shade always move towards the shade. This makes sense, since we know chloroplasts avoid bright light. However, if a chloroplast is completely inside the light it will not move away, but it will stay in place, even though the high-intensity light is bad for it. This suggested to us that it's not the intensity of the light that drives the motion of chloroplasts, but the difference in intensity between two places, so a light gradient. Secondly, we really wanted to understand why the chloroplasts cluster inside the cell after the light is turned on. We knew from observation made by Nico that chloroplasts can actually give shade to each other by being on top of each other. We thought that this shading effect could lead to some sort of attraction between the particles and that this could be what drives the fun clustering we saw in our videos. We decided to use our model to test both of these hypotheses.

Forward a few months later to the point were I managed to have a working code[15]. We did some simulations of a single particle and compared these to actual experiments done by other people in the past. We found that light gradients do indeed seem to drive the motion of chloroplasts. Hypothesis one out of two was correct! Encouraged by this we thought, let's move on to looking at the effect of the shading between particles. To do that we did simulations with eighty particles. We chose eighty since that is a similar amount to the number of chloroplasts inside an *Elodea densa* cell[16]. With these simulations we found that the shadow effect between particles does lead to them forming clusters, but our simulated clusters were not nearly as fun and

---

[14]Most of these were biology papers. I have a very strong opinion on those as you could probably tell from the introduction.

[15]With the help of my best friend and amazingly smart human being Renske. I am eternally grateful to her, she essentially wrote half of the code for this model.

[16]We know that because we counted them. I know, science is riveting sometimes.

hyperactive as the ones formed by actual chloroplasts. The clusters of particles in our simulations didn't move at all, while the clusters of chloroplasts moved quite a bit inside the cell. From this we concluded that, although the shading between chloroplasts is necessary for these particles to cluster, it is not the only effect involved in the movement we see in the experiments. Consequently, other physical effects will need to be added to our model in order to simulate the motion of chloroplasts accurately.

You might be thinking that this is a bit of a let down, but no. Even though our model does not reproduce chloroplast motion completely, it is still very nice and useful. It could be used in future research as a basis to build on and add more physical effects for us to test what *does* lead to the motion we see in experiments. The model we proposed here could even serve as the basis for a model that can reproduce all chloroplast motion! This is quite some time away though. My hope is that this thesis and the article we wrote about the dim-light adapted chloroplasts [33] lead to more people being interested in the motion of chloroplasts. Continuing to study this system could teach us a lot, not only about the physics of it, but also from a biological perspective. However, this will require biologists and physicists to work together so that, maybe, through the power of friendship, we can uncover new details about our wonderful world.

Acknowledgements

First, I would like to thank my supervisors Nico Schramma and Mazi Jalaal. I had a lot of fun during this project and I feel very lucky to have been able to work with them. Having them as my supervisors has been an amazing experience. I especially want to thank Nico for all his time, help and guidance. Thank you for pushing me every time to think outside the box and become a better researcher.

Furthermore, I want to thank my amazing friend Renske Wierda for essentially coding half of my simulations, making the beautiful front page of this thesis, for being an amazing editor and for always taking the time to listen to me rant about my problems, even though she has no idea what I am talking about half of the time. I also want to thank my good friend Tim Veenstra for all the fun work and lunch sessions we had together on Mondays, which helped to keep motivated, and for proofreading my thesis (multiple times) even though he was very busy with his own research. I really could not have done this project without either of these lovely human beings.

Additionally, I thank my mum for helping me understand and incorporate biological knowledge into this thesis in a seamless way. Without her there would have been many mistakes in the biology section of this thesis.

Lastly, I would like to thank Jazza, Sarah Renae Clark, Moriah Elizabth, Rae Dizzle and Temi Danso for keeping me entertained and helping me relax during the past year. I would also like to thank the LGBTQIA+ community for inventing the rainbow, which I used so many times throughout this thesis.